Advanced Battery Technologies

Advanced Battery Technologies: New Applications and Management Systems

Editors

Manuela González
David Anseán

MDPI • Basel • Beijing • Wuhan • Barcelona • Belgrade • Manchester • Tokyo • Cluj • Tianjin

Editors
Manuela González
University of Oviedo
Spain

David Anseán
University of Oviedo
Spain

Editorial Office
MDPI
St. Alban-Anlage 66
4052 Basel, Switzerland

This is a reprint of articles from the Special Issue published online in the open access journal *Electronics* (ISSN 2079-9292) (available at: https://www.mdpi.com/journal/electronics/special_issues/abtbms).

For citation purposes, cite each article independently as indicated on the article page online and as indicated below:

LastName, A.A.; LastName, B.B.; LastName, C.C. Article Title. *Journal Name* **Year**, *Volume Number*, Page Range.

ISBN 978-3-0365-0922-8 (Hbk)
ISBN 978-3-0365-0923-5 (PDF)

Cover image courtesy of David Anseán.

© 2021 by the authors. Articles in this book are Open Access and distributed under the Creative Commons Attribution (CC BY) license, which allows users to download, copy and build upon published articles, as long as the author and publisher are properly credited, which ensures maximum dissemination and a wider impact of our publications.

The book as a whole is distributed by MDPI under the terms and conditions of the Creative Commons license CC BY-NC-ND.

Contents

About the Editors . . . . **vii**

Preface to "Advanced Battery Technologies: New Applications and Management Systems" . . **ix**

Matthieu Dubarry and George Baure
Perspective on Commercial Li-ion Battery Testing, Best Practices for Simple and Effective Protocols
Reprinted from: *Electronics* **2020**, *9*, 152, doi:10.3390/electronics9010152 . . . . **1**

Pavol Spanik, Michal Frivaldsky, Juraj Adamec and Matus Danko
Battery Charging Procedure Proposal Including Regeneration of Short-Circuited and Deeply Discharged LiFePO$_4$ Traction Batteries
Reprinted from: *Electronics* **2020**, *9*, 929, doi:10.3390/electronics9060929 . . . . **17**

Qi Zhang, Yan Li, Yunlong Shang, Bin Duan, Naxin Cui and Chenghui Zhang
A Fractional-Order Kinetic Battery Model of Lithium-Ion Batteries Considering a Nonlinear Capacity
Reprinted from: *Electronics* **2019**, *8*, 394, doi:10.3390/electronics8040394 . . . . **33**

Yidan Xu, Minghui Hu, Chunyun Fu, Kaibin Cao, Zhong Su and Zhong Yang
State of Charge Estimation for Lithium-Ion Batteries Based on Temperature-Dependent Second-Order RC Model
Reprinted from: *Electronics* **2019**, *8*, 1012, doi:10.3390/electronics8091012 . . . . **49**

Yongliang Zheng, Feng He and Wenliang Wang
A Method to Identify Lithium Battery Parameters and Estimate SOC Based on Different Temperatures and Driving Conditions
Reprinted from: *Electronics* **2019**, *8*, 1391, doi:10.3390/electronics8121391 . . . . **69**

Jiechao Lv, Baochen Jiang, Xiaoli Wang, Yirong Liu and Yucheng Fu
Estimation of the State of Charge of Lithium Batteries Based on Adaptive Unscented Kalman Filter Algorithm
Reprinted from: *Electronics* **2020**, *9*, 1425, doi:10.3390/electronics9091425 . . . . **83**

Yi Wu, Youren Wang, Winco K. C. Yung and Michael Pecht
Ultrasonic Health Monitoring of Lithium-Ion Batteries
Reprinted from: *Electronics* **2019**, *8*, 751, doi:10.3390/electronics8070751 . . . . **105**

Christian Brañas, Juan C. Viera, Francisco J. Azcondo, Rosario Casanueva, Manuela Gonzalez and Francisco J. Díaz
Battery Charger Based on a Resonant Converter for High-Power LiFePO$_4$ Batteries
Reprinted from: *Electronics* **2021**, *10*, 266, doi:10.3390/electronics10030266 . . . . **121**

Fengdong Shi and Dawei Song
A Novel High-Efficiency Double-Input Bidirectional DC/DC Converter for Battery Cell-Voltage Equalizer with Flyback Transformer
Reprinted from: *Electronics* **2019**, *8*, 1426, doi:10.3390/electronics8121426 . . . . **141**

Sadam Hussain, Muhammad Umair Ali, Sarvar Hussain Nengroo, Imran Khan, Muhammad Ishfaq and Hee-Je Kim
Semiactive Hybrid Energy Management System: A Solution for Electric Wheelchairs
Reprinted from: *Electronics* **2019**, *8*, 345, doi:10.3390/electronics8030345 . . . . **159**

About the Editors

Manuela González (IEEE Senior Member) received her MSc and PhD (with honors) in Electrical Engineering from the University of Oviedo, Spain, in 1992 and 1998, respectively. From 1993, she has been working at the Electronics and Electrical Engineering Department; since 2000, she has been an Associate Professor there. From 1998 to present, she has been the technical leader of the "Battery and New Energy Storage Systems" researching group. The multidisciplinary team (PhD Electrical and Electronics Engineers, PhD Computer Science Engineers and PhD in Chemistry) focuses on the research of the electrical and chemical characterization of batteries, and on the design of battery management systems (BMS), including battery modeling, fast-charging, state-of-charge (SoC) and state-of-health (SoH) analysis. In recent years, her research has been focused on novel lithium-ion battery technologies (LFP, NMC, LTO, etc.) for electric vehicles (EVs) and batter energy storage systems applications. In the battery research line, she has led more than 20 R&D projects and collaborated in 8 R&D projects that have been developed with the collaboration and/or financing of companies in the field of batteries, portable applications, electric traction, EV chargers, etc. She has 25 publications with relative quality indexes, and more than 50 papers in international congresses, in addition to 5 invited conferences. Moreover, she has participated in the organization of international congresses sponsored or co-sponsored by the IEEE. In the associated research line of electronic instrumentation, she has collaborated on 11 R&D projects, and she has 18 publications with relative quality index and 19 papers in international congresses.

David Anseán is an Assistant Professor at the Department of Electrical Engineering, University of Oviedo (Spain). Dr Anseán received his MEng degree from the University of Granada, (Spain) in 2007, and his PhD (with honors) from the University of Oviedo in 2015, both in electronics engineering. Before pursuing his PhD, he gained international industry experience in Basingstoke, UK, and Berkeley, CA, USA. As a doctoral student, he was the recipient of a research fellowship stay at the Electrochemical Power Systems Laboratory, at the University of Hawaii, USA, which he later joined as a Postdoctoral Fellow, to work in Dr. Dubarry's group on advanced diagnosis and prognosis techniques of lithium-ion batteries. Since 2016, he has been an Assistant Professor at the University of Oviedo, where he is the instructor of undergraduate and graduate courses including Power Electronics, Digital Integrated Circuits, Electronics Instrumentation, and Embedded Systems. His research interests include lithium-ion battery degradation mechanisms analysis via non-invasive methods, battery testing and characterization, and the design of battery fast charging protocols. In 2018 and in 2019, he was the recipient of Visiting Scholar Research Fellowships and joined the Institute for Power Electronics and Electrical Drives (ISEA) at RWTH Aachen University (Germany), and the Electrochemical Power Systems Laboratory, at the University of Hawaii (USA), respectively.

Preface to "Advanced Battery Technologies: New Applications and Management Systems"

1. Introduction

Lithium-ion batteries (LIBs) are ubiquitous in our modern society. We can find them in every type of electronic devices, with common examples including mobile phones, laptops, smartwatches, and digital cameras. But the range of applications of LIBs has expanded vastly in the last years. Nowadays, LIBs are in fact the core power system in electric mobility (i.e., Electric and Hybrid Vehicles, Electric Trains, Electric Bikes and Electric Vessels), play a key role in large-scale energy storage systems for renewables, and remain essential in niche applications, such as aerospace industry and biomedical instruments (pacemakers, defibrillators, etc.). We can therefore confirm that LIBs are a key-technology in present and future engineering systems.

Despite the achievements and applicability of LIBs, there are several features within this technology that require further improvements. For instance, measuring the inner state of charge (SOC) and state of health (SOH) of a LIB remains challenging today. Advances to better evaluate these two figures of merit (i.e., SOC and SOH) are key to improve reliability and controllability in the increasingly complex LIB systems. From a system-level perspective, the continued improvements in overall system-efficiency are also critical. This area is usually explored via Power Electronics. Another key feature, incidentally, less taken into consideration, is related to laboratory battery testing. This is indeed an integral part of any study dealing with LIBs laboratory-based research. These are just a few examples of key features to be addressed in LIBs. Curious readers are encouraged to refer to the state-of-the-art literature of LIBs for further study.

We must also emphasize that the understanding of LIBs encompasses diverse disciplines that not only includes Electronic Engineering, but also Control and Computer Engineering, Instrumentations and Measurement Engineering, Material Science, Electrochemistry, or Data Science. Therefore, to overall improve LIB performance and control, a multidisciplinary team (and set of skills) is required. This aspect is central, yet it was often not taken into consideration. Fortunately, nowadays multidisciplinary teams are more common in industry, research, and academia.

In this Special Issue, we are pleased to present 10 high-quality papers that cover the above-mentioned matters, with a diversity of focus areas of LIBs that include:

1. Battery testing methodologies and operation procedures [1,2]
2. Battery state of charge monitoring [3–6]
3. Battery State of health monitoring [7]
4. Power Electronics applications in Lithium-ion batteries [8–10]

In Figure 1, we summarize the organization of this Special Issue to facilitate its study. The sequence of topics was selected from lower-level, laboratory-oriented topics (1. Battery testing), inner, key figures of merit in LIBs (2. State of Charge and 3. State of Health), to final, system-level applications (4. Power Electronics). Similarly, Figure 1 also includes the paper references for quick access, together with the main areas of expertise within the selected topics to help readers jump straight into their area of interest.

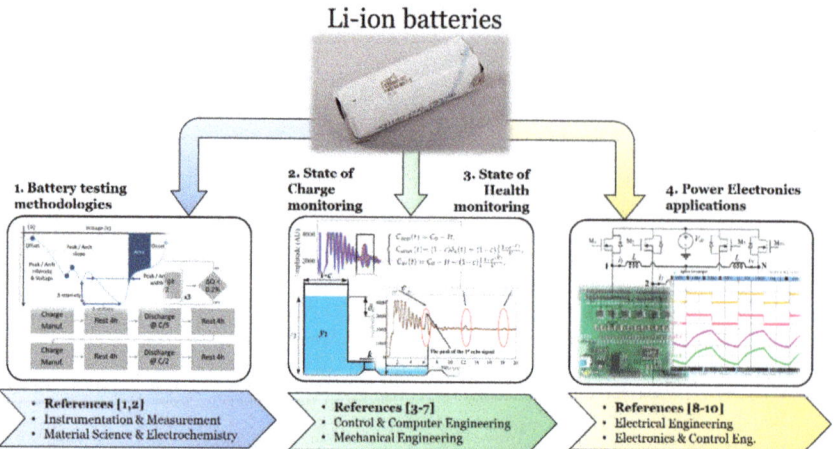

Figure 1. Summary of the topics and disciplines covered in this Special Issue. All pictures and graphs are taken from the research papers compiled in this book.

Another key aspect to highlight in this Special Issue is the variety of experiments and experimental set-ups presented. The works include the assembly of prototypes in printed circuit boards, battery testing, experimentation and measurements, and the validation of the developed mathematical models via experiments. We believe that this variety of conducting tests is to be valued by the readers of this Special Issue.

Finally, it is also worth noting that one paper was selected as a feature paper for the Special Issue:

M. Dubarry, G. Baure, Perspective on commercial Li-ion battery testing, best practices for simple and effective protocols, Electron. 9 (2020). doi:10.3390/electronics9010152.

and it is also worth noting that the work reported in another paper was also presented in the form of a patent:

Q. Zhang, Y. Li, Y. Shang, B. Duan, N. Cui, C. Zhang, A fractional-order kinetic battery model of lithium-ion batteries considering a nonlinear capacity, Electron. 8 (2019). doi:10.3390/electronics8040394.

2. The Present Special Issue

2.1. Battery Testing Methodologies and Operation Procedures

Laboratory testing is expensive, time-demanding, and a much more complex task than it appears. Properly defined testing protocols and standard operating procedures are key to further improve the understanding of LIBs in many respects: study and validating modeling, development of control algorithms, battery performance, or safety. In [1], the testing approach used at the Hawaii Natural Energy Institute (HNEI) and developed for over 15 years is described in detail. The HNEI has been in the forefront of the development of methodologies to improve non-intrusive characterization of commercial lithium-ion cells to extract maximum relevant information from minimum amount of testing and instrumentation. The paper will help engineers and researchers, both in industry and academia to improve their battery testing capabilities and validation studies. In [2], an investigation of the recovery of damaged batteries via testing is presented. For this purpose, experimental set-up for automated system integrating proposed recovery methods were realized and analyzed, attaining a high recovery-rate out of damaged cells.

2.2. Battery State of Charge Monitoring

State of charge (SOC) accurate estimation is one of the most important functions in a battery management system. Several methodologies for SOC estimation are generally explored, including adaptive filter algorithms, learning algorithms, nonlinear observers or hybrid methods, to name a few. In [3], a fractional-order kinetic model is developed to estimate the available capacity of a battery, taking into consideration the battery dynamics of the electrochemical materials and properties. In [4], a temperature-dependent, second order RC equivalent circuit is developed together with a dual Kalman filter algorithm. The estimation results were validated from experiments, attaining high accuracy in the results. Further improvements of the unscented Kalman filter were presented in [6], by developing an adaptive unscented Kalman filter. The proposed algorithm reduces the system noise and observation noise during SOC estimation, which leads to an improve of accuracy than the dual Kalman filter algorithm.

2.3. Battery State of Health Monitoring

Due to the non-linear physiochemical nature of LIBs, identify the internal changes that lead to battery degradation and failure is a complex task. In [7], a novel health monitoring method based on ultrasonic techniques is presented. The developed method offers a significant improvement over the state-of-the-art ultrasonic techniques, in terms of providing an early indication of sudden battery failure. In addition, the monitoring method can be applied to batteries during their operation by integrating simple and small equipment (a pulser-receiver module and a piezoelectric transducer) into the existing battery management system.

2.4. Power Electronics Applications in Lithium-Ion Batteries

System-level management of LIBs is essential to improve performance, protect the batteries and maximize the lifespan of the system. One area of key interest within the system-level management of LIBs is the charging strategy. Improving the charging capabilities of LIBs involves both the development of the charging protocol and the electronics to accomplish the charging scheme. In [8], a multiphase resonant converter for high-capacity, 48 V LiFePO4 battery module is presented. The designed converter offers high reliability throughout the charging process, improving the charging capabilities of the tested pack. Another application of Power Electronics in LIBs is presented in [9]. Here, a bidirectional direct current-direct current equalization structure is developed to equalize batteries within a battery pack, without using an external energy buffer. With the proposed architecture, a good balance of performance is attained in speed, design cost and volume. To close this Special Issue, a hybrid energy management system is presented in [10]. The system contained a LIB and supercapacitor connected to a DC bus via bidirectional DC-DC converter. The designed prototype validates the proposed architecture, attaining improved values in the area of energy efficiency and charging time.

3. Concluding Remarks

The Guest Editors were pleased with the quality and breadth of the accepted papers. Looking to the future, we believe all research works enclosed in this Special Issue will promote further research in Lithium-ion batteries from a multidisciplinary perspective.

Author Contributions: The authors worked together and contributed equally during the editorial process of this Special Issue.

Funding: This research received no external funding.

Acknowledgments: The Guest Editors thank all of the authors for their excellent contributions to this Special Issue. We also thank the reviewers for their dedication and suggestions to improve each of the papers. We finally thank the Editorial Board of MDPI's Electronics for allowing us to be Guest Editors for this Special Issue, and to the Electronics Editorial Office for their guidance, dedication, and support.

Conflicts of Interest: The authors declare no conflict of interest.

Manuela González and David Anseán

Guest Editors

References

1. Dubarry, M.; Baure, G. Perspective on commercial Li-ion battery testing, best practices for simple and effective protocols. *Electronics* **2020**, *9*, 152, doi:10.3390/electronics9010152.
2. Spanik, P.; Frivaldsky, M.; Adamec, J.; Danko, M. Battery Charging Proposal Including Regeneration of Short-Circuited and Deeply Discharged LiFePO 4 Traction Batteries. *Electronics* **2020**, *9*, 929, doi:10.3390/electronics9060929.
3. Zhang, Q.; Li, Y.; Shang, Y.; Duan, B.; Cui, N.; Zhang, C. A fractional-order kinetic battery model of lithium-ion batteries considering a nonlinear capacity. *Electronics* **2019**, *8*, 394, doi:10.3390/electronics8040394.
4. Xu, Y.; Hu, M.; Fu, C.; Cao, K.; Su, Z.; Yang, Z. State of charge estimation for lithium-ion batteries based on temperature-dependent second-order RC model. *Electronics* **2019**, *8*, 1012, doi:10.3390/electronics8091012.
5. Zheng, Y.; He, F.; Wang, W. A method to identify lithium battery parameters and estimate SOC based on different temperatures and driving conditions. *Electronics* **2019**, *8*, 1391, doi:10.3390/electronics8121391.
6. Lv, J.; Jiang, B.; Wang, X.; Liu, Y.; Fu, Y. Estimation of the state of charge of lithium batteries based on adaptive unscented Kalman filter algorithm. *Electronics* **2020**, *9*, 1425. doi:10.3390/electronics9091425.
7. Wu, Y.; Wang, Y.; Yung, W.K.C.; Pecht, M. Ultrasonic health monitoring of lithium-ion batteries. *Electronics* **2019**, *8*, 751, doi:10.3390/electronics8070751.
8. Brañas, C.; Viera, J.C.; Azcondo, F.J.; Casanueva, R.; Gonzalez, M.; Díaz, F.J. Battery charger based on a resonant converter for high-power lifepo4 batteries. *Electronics* **2021**, *10*, 266, doi:10.3390/electronics10030266.
9. Shi, F.; Song, D. A novel high-efficiency double-input bidirectional DC/DC converter for battery cell-voltage equalizer with flyback transformer. *Electronics* **2019**, *8*, 1426, doi:10.3390/electronics8121426.
10. Hussain, S.; Ali, M.U.; Nengroo, S.H.; Khan, I.; Ishfaq, M.; Kim, H.J. Semiactive hybrid energy management system: A solution for electric wheelchairs. *Electronics* **2019**, *8*, 345, doi:10.3390/electronics8030345.

Perspective

Perspective on Commercial Li-ion Battery Testing, Best Practices for Simple and Effective Protocols

Matthieu Dubarry * and George Baure

Hawaii Natural Energy Institute, University of Hawaii, 1680 East West Road, POST 109, Honolulu, HI 96815, USA; gbaure@hawaii.edu
* Correspondence: matthieu.dubarry@gmail.com

Received: 14 December 2019; Accepted: 8 January 2020; Published: 14 January 2020

Abstract: Validation is an integral part of any study dealing with modeling or development of new control algorithms for lithium ion batteries. Without proper validation, the impact of a study could be drastically reduced. In a perfect world, validation should involve testing in deployed systems, but it is often unpractical and costly. As a result, validation is more often conducted on single cells under control laboratory conditions. Laboratory testing is a complex task, and improper implementation could lead to fallacious results. Although common practice in open literature, the protocols used are usually too quickly detailed and important details are left out. This work intends to fully describe, explain, and exemplify a simple step-by-step single apparatus methodology for commercial battery testing in order to facilitate and standardize validation studies.

Keywords: commercial Li-ion testing; RPT; CtcV; cell-to-cell variations

1. Introduction

Today's world relies more and more on energy storage technologies and, with several government incentives for larger integration of zero-emission electricity storage in electromobility and stationary applications, the demand will keep increasing in the future [1]. To match this demand, battery technology must improve year after year. To be more than incremental, such improvement could take the form of disruptive battery technologies [2,3] and, equally as important, the form of improved battery management systems with innovative control strategies to enable more efficient and safer battery packs. The latter topic is attracting enormous amount a research and a wide variety of algorithms have been proposed in recent years [4–8] for state-of-charge (SOC) and state-of-health (SOH) tracking. For all these studies, there is a dire need for experimental validation so that the effectiveness of the proposed methodology can be demonstrated. This is often neglected, and some studies, as promising as they could be, are disregarded because the work was not properly validated in the laboratory.

Laboratory testing is an expensive and complex task, especially when trying to replicate the behavior of large deployed battery packs such as the ones in electric vehicles or grid storage systems. Laboratory results can become non-significant if cells are not handled and characterized properly. In regard to characterization, laboratory testing at scale is often not possible because of logistical and cost limitations, most of the testing must then be performed at a much smaller scale and under slightly different conditions. This raises concerns about the presumptions that the tested cells are representative of the batch, the duty cycle is relevant to the application, and the pack behaves similarly to single cells at scale. To address the representativity issue, cell-to-cell variations need to be studied and quantified. To address the relevance issue, duty cycles must be illustrative of application data. To address the degradation and state-of-health issue, cells need to be periodically characterized in a non-intrusive and non-destructive way.

The purpose of this publication is to complement [9,10] and describe, in more detail than typical publications, a testing strategy to address all these issues together. This is meant to provide newcomers and non-battery specialists some details, definitions, and explanations on how to perform simple and effective battery testing in order to improve validation studies. This publication does not aim to compare different approaches to testing protocols nor discuss the complex question of battery SOH, such discussions can be found in [9,10], respectively, but is intended to describe the Hawaii Natural Energy Institute (HNEI) methodology. In the past decade and a half, HNEI has been in the forefront of the development of methodologies to improve non-intrusive characterization of commercial lithium ion cells to extract maximum relevant information from minimum amount of testing and instrumentation. In that timeframe, we tested over 1000 commercial cells and published upwards of 50 highly cited publications on commercial battery testing and modeling.

The HNEI testing strategy consists of five elements, Figure 1. First, a preparation step must be executed to assure the cells are properly installed, the testers are accurately calibrated, and the compulsory safety precautions are strictly implemented. Second, a formation protocol must be performed to verify the cell quality relative to the batch. Third, a reference performance test (RPT) should be completed at regular intervals to assess the evolution of battery performance over time. Fourth, a repetitive duty cycle is required to mimic battery usage for a given application. Lastly, an end-of-test evaluation is undertaken to provide a detailed characterization of cell performance at the end of life, which includes a final RPT and, if deemed necessary, some post-mortem analyses [11,12].

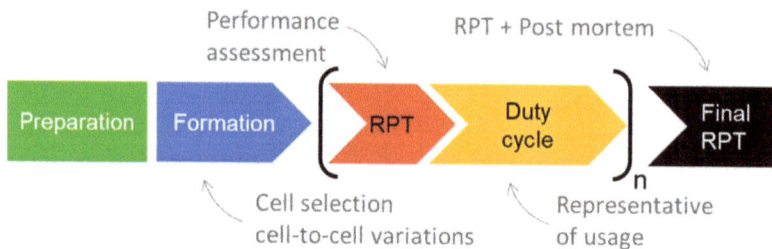

Figure 1. Testing sequence within proposed protocols.

2. Methods and Discussion

2.1. Test Preparation

A recent study by Taylor et al. [13] showed that, without careful consideration, slight differences in the experimental setup could induce up to 4% difference in experimental results on similar cells. This error has environmental and procedural origins [13]. The environmental errors could stem from the ambient temperature and humidity; the equipment calibration, accuracy, and resolution; or the different manufacturing tolerances of the equipment and battery used. Procedural errors are induced during the testing, and include many circumstances such as moving a sample, loosening a connection, or not allowing batteries to acclimate to a new temperature properly [13].

Such a high error could drastically influence the conclusions of a study. Thus, steps should be taken to minimize it. First, battery tester calibrations should be verified. Second, battery holder contacts must be cleaned to avoid the possible impact of oxidation. Third, the placement of the connector cables and eventual spacers must be consistent. The measurement cables must be as close as possible to the battery tabs. In case of 4-point connection (2 for voltage, 2 for current), it is recommended that the current and voltage cables do not touch each other [13]. Fourth, the torque on the connections must be carefully controlled and kept constant. Taylor et al. [13] found the optimal torque to be 12.5 Nm for their connectors. If possible, all cells belonging to one set of experiments should be tested on the same machine and in the same temperature chamber if temperature is not a variable. Finally, thermocouples must be added, and their placement should also be consistent (same position, same amount of tape, etc.). With those steps, the experimental error should be limited to below 1% [13].

Another essential aspect to take into consideration while preparing an experiment is safety. Even if no abusive testing is performed, failure is always an option. Modern batteries pack a lot of energy as 55Ah worth of batteries is equivalent to the energy of a hand grenade (150 g of TNT) [14]. When first received, the batteries should be unpacked under a fume hood to prevent any exposure from potential electrolyte leakage during transport [14]. They should then be thoroughly checked for any physical damage, leak, or defect. Defective batteries should be disposed according to local health and environmental safety office recommendations. For storage, batteries should be discharged a low to mid SOCs, vacuum sealed in non-metallized plastic bag, and frozen to −27 °C in a commercial freezer. This is because low temperatures and SOCs were shown effective to prevent impact of calendar aging for all the major commercial Li-ion chemistries [15]. A plan should also be in place in case batteries undergo thermal runaway during testing. A discussion on thermal runaway is out of the scope of this publication and interested readers should refer to [16–19]. To maximize safety, all cells should be tested in temperature chambers with significant exhaust ventilation to evacuate fumes quickly in case of failure. Moreover, temperature should always be monitored as it is an excellent indicator of failure [16–19]. If cell temperature exceeds 80 °C, all testing should be stopped, and the temperature monitored closely for the next hour. Electrolyte decomposition is an exothermic reaction and if happening, the temperature of the cell will continue to rise. Therefore, if the cell temperature returns to room temperature quickly, the risk of dramatic failure is low. Some cooling aid could be applied to the cells in the form of ice packs [14]. Once cooled, and as precaution, the cell should be transferred to a sand-filled bucket [14], if possible, made of earthenware. Sand offers the advantage of acting like a sponge for any leak of boiling electrolyte without the risk of burning or melting. The container must be kept sealed, outdoor but protected from weather, and temperature monitored for 72 h. If the temperature keeps increasing, the risk of failure is high. Extreme caution must be exercised as venting or explosion is possible at any moment without notice. Standard operating procedures and proper training should be installed to be able to handle such events quickly and safely. These procedures could include aggressive cooling solutions such as ice packs, dry ice, and CO_2 fire extinguishers and the placement of fire blankets over the overheated cell and on the adjacent cells to prevent propagation. Personal protective equipment such as heat resistant gloves, a fire-retardant lab-coat, and a full-face respirator, must be used in all cases. An adapted first aid kit must also be in close proximity [14].

2.2. Formation

Before the start of any cycle-life evaluation, it is extremely important to identify and quantify the nature of cell-to-cell variations within a batch of cells [20–25]. A discussion on their origins is out of the scope of this publication and interested readers should refer to an article by Rumpf et al. [26]. The results of the formation tests are not reported often enough. Several strategies are available in the literature from C/3 discharge and 50% SOC resistance test [27] to protocols such as HNEI's initial conditioning characterization test (ICCT) [28–31]. Lasting less than a week, the ICCT consists only of C/2 and C/5 cycles, Figure 2, and serves two purposes. The first is to verify that the cells are working as they should be and that the solid electrolyte interphase layers are properly formed [32]. The second is to calculate the three parameters that were shown to be critical in determining the manufacturing variability in a batch of cells [28]. For the ICCT and all the other protocols in this publication, data collection must be controlled to ensure enough information is gathered while limiting file size. We recommend some variable time steps aiming for 2000 points per step. For the C/5 step that is supposed to last around 5 h, 1 point should be recorded every 9 s. For the C/2 cycle, that measurement rate is accelerated to 1 point every 3.6 s.

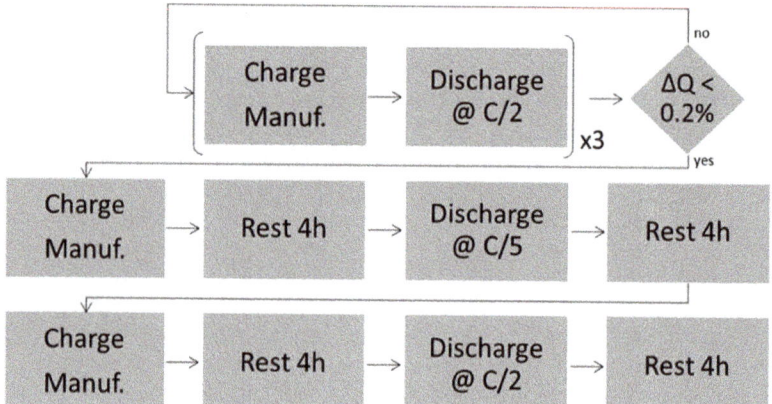

Figure 2. Initial conditioning characterization test (ICCT) formation test protocol.

Figure 2 details the ICCT protocol. The aim of the first step is capacity stabilization. It is recommended to start by performing a few charge and discharge cycles (up to 6) to ensure the SEI layer on the negative electrode is properly formed. Once the capacity is stabilized (less than 0.2% difference between two consecutive cycles), the second step can be started. The second step consists of C/2 (discharge in 2 h) and C/5 (discharge in 5 h) discharges with 4-h rests and the manufacturer-recommended charging protocol.

The capacities and rest cell voltages (RCV), i.e., the voltage measured at the end of a resting period, measured throughout this test are used to calculate the three attributes that are critical in determining the manufacturing variability in a batch of cells. In order to fully characterize a batch of cells, it is important to fully compare the capacity vs. rate relationship for each cell. This relationship is not straightforward and it can be divided into 3 sections, Figure 3a.

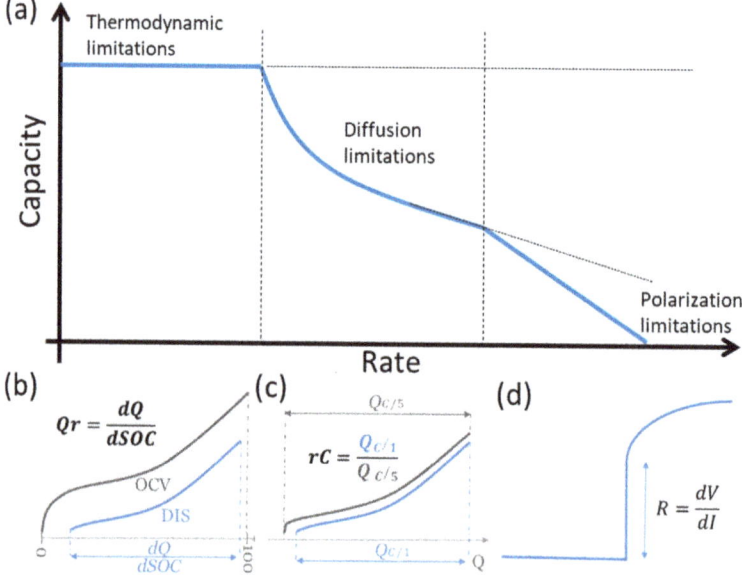

Figure 3. (a) Capacity vs. rate relationship and (b–d) the three attributes that quantify cell-to-cell variations.

In the first section at low rates, the capacity is constant because it is only limited by the amount of lithium that can be exchanged. This corresponds to the cell maximum capacity and it can be characterized by calculating the capacity ration (Qr, in mAh/% SOC, Figure 3b). The term capacity ration refers to the capacity (Ah) obtained for each one percent of SOC. RCV measurements at the beginning of discharge (BOD) and the end of discharge (EOD) are used to derive a SOC range by interpolation of the maximum and minimum SOCs (e.g., 99.7%–3.2%) from an open circuit voltage (OCV) vs. SOC curve. The capacity ration is then calculated by dividing the capacity returned during discharge by the SOC range variation. The maximum capacity corresponds to 100 x Qr, the capacity for 100% SOC.

In the second section at medium rates, the capacity becomes also limited by diffusion [33] and starts decreasing with rate following a power law. This section can be characterized either by the Peukert coefficient [34] or the rate capability (rC, Figure 3c). The Peukert coefficient [34] can be calculated by fitting the data in the middle section to a $C = I^n t$ equation where n is the Peukert coefficient, I the current, and t the nominal discharge time for a specific C-rate. The rate capability represents a cell's ability to deliver capacity when the discharge rate increases. It can be calculated by dividing the capacities at C/2 by the C/5 ones. Both parameters are unitless and, with only two rates tested in the ICCT, rC if often more appropriate.

In the last section, at high rates, in addition to being limited by the amount of lithium and diffusion, the capacity starts to be affected by polarization pushing some capacity outside of the potential window. This can be characterized by measuring the ohmic resistance (R, in Ohms, Figure 3d) [28]. The ohmic resistance represents the contact resistance of the cell in the circuit and the conductive resistance of the cell (which primarily comes from the electrolyte). Although several methods could be used to estimate the resistance, such as electrochemical impedance spectroscopy (EIS), the resistance estimation can be obtained simply by using the initial voltage drop associated with the C/2 and C/5 discharges. The method is based on the linear regime of the Tafel behavior [35] which, for small currents, shares some formal similarity with Ohm's law. Figure 4a presents the calculation process that was described previously in [36]. If the cells were previously charged to the exact same SOC prior to the C/5 and C/2 discharges, i.e., that the RCVs were similar, the initial IR drop can be used to determine the resistance. To calculate the resistance, the measured voltage must be plotted as a function of rate or current. The slope of the curve is the resistance, normalized or not. It must be noted that if the data are gathered at different time steps for different rates, priority must be set on selecting points with similar elapsed time after the application of current. This approach is also valid to characterize the resistance evolution with temperature and aging.

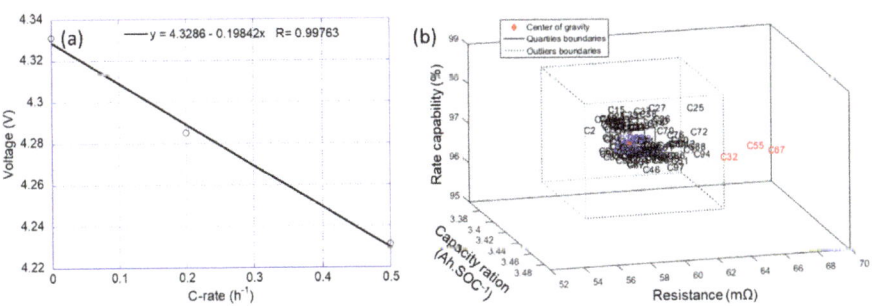

Figure 4. (**a**) Initial voltage vs. rate for the resistance calculation and (**b**) cell-to-cell variation representation in a 3D space adapted from [29].

Once the three attributes are calculated for all the cells in a batch, the statistical analysis of the cell-to-cell variations can be performed, and the normality of the three attributes examined. A convenient way to report the data is to represent all three attributes in a 3D space of which an example is presented in Figure 4b. To help visualize the cell-to-cell variations better, a couple of visual cues can be added to the plot. First, a full-line rectangle, in which the boundaries correspond to the lower and upper quartiles (i.e., the 75th and 25th percentiles of the data) for the three attributes of cell-to-cell variations. Cells located within this rectangle, color-coded in blue, are the closest to batch center of gravity (the median values for all three attributes) and can be considered the core of the batch. A second dotted rectangle, three times larger than the first one, was also added. Three times the interquartile range is often statistically considered as the outliers' boundaries. Cells located outside of the outliers' boundaries were color-coded in red and are excluded from further experimentation. In the example shown in Figure 4b, most cells were consistent. However, there were three cells that exhibited resistances higher than normal and, therefore, their use should be discontinued.

2.3. Reference Performance Test

The RPT needs to be performed on a regular interval and at a constant temperature, typically monthly or every 100 full equivalent cycles and at room temperature, to quantitatively assess battery performance and SOH. SOH is gauged from the quantification of the degradation modes, the loss of active material, loss of reactant, and kinetic degradation. A discussion on SOH is out of the scope of this paper and can be found in [10]. The RPT needs to provide as much information as possible on SOH without being intrusive [37]. Therefore, the test must be short and not stress the cells, but it needs to characterize the thermodynamic and kinetic changes. The proposed HNEI RPT protocol is detailed in Figure 5. It consists of three steps: Conditioning, low-rate cycling, and nominal rate cycling. The conditioning step ensures that the cells are fully charged before the start of the low-rate cycle. This is accomplished by performing a standard charge with an additional constant current step typically at C/50 (discharge in 50 h). The second step is a low-rate cycle, typically C/25 (discharge in 25 h) to assess thermodynamic aspects. The third step is a nominal rate cycle, typically C/2 (discharge in 2 h) to assess kinetic aspects. Rate for this step can be adapted to the situation. For example, higher rates can be applied for high-power cells; while lower rates are used for high-energy cells. The caveat being that this rate must be less aggressive than the rate used in the duty cycling. The RPT protocol is limited to two cycles to minimize its toll on the cell SOH but additional rates or procedures (constant power, electrochemical impedance spectroscopy, full OCV vs. SOC test, different temperatures) [27,38,39] could be added if necessary. This is often done in the literature for the first RPT only. All regimes are performed at constant current up to the cutoff voltage with the addition of a residual capacity step at C/50 in between two rests. This residual capacity step is primarily used to assure that both charges and discharges start from the same SOC independently of the previous rate. It is also used to calculate the maximum capacity the cell can deliver.

The data gathered during the RPT test allow the execution of several analyses. First, the same calculation as that of the ICCT test can be undertaken, specifically, the capacity ration, rate capability, and resistance can be measured. The main difference is that in this case, the capacity ration can be deciphered directly by adding the capacity measured during the residual capacity measurements at C/50 to the recorded capacities to obtain the cell maximum capacity to be divided by 100. Values calculated for the two rates should be similar and close to the one inferred during the ICCT test. With only two rates tested during the proposed RPT, the rC method is still recommended over the Peukert method to characterize rate capability. If more rates are tested, the Peukert curve (capacity vs. rate) could be used as long as the rates are within the section with diffusion limitation (Figure 3a).

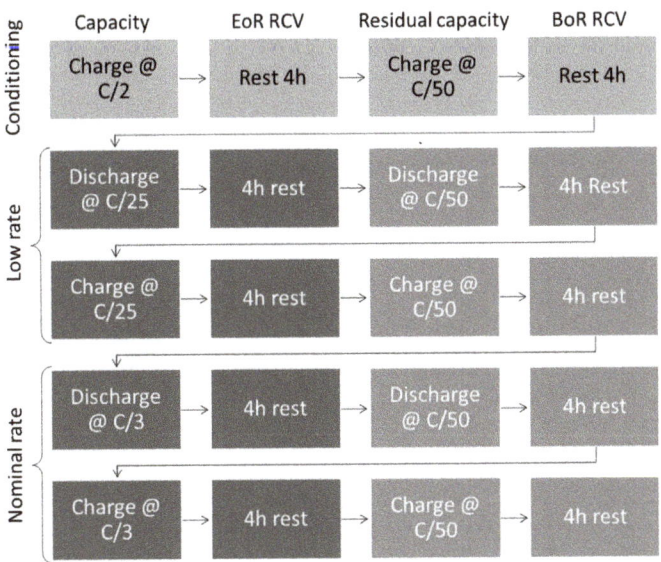

Figure 5. The reference performance test (RPT) test protocol. EoR stands for end of regime and BoR stands for beginning of regime.

The second set of information to be gathered from the RPT is the OCV vs. SOC relationship at the current SOH. A discussion on OCV is out of the scope of this paper and can be found in [9,40]. The OCV vs. SOC relationship can vary a lot with aging and, since gathering a proper OCV vs. SOC curve can take weeks of testing [9], it is not feasible to repeat the process on every cell at each RPT. A solution is to calculate a pseudo-OCV curve at each RPT. It has to be stated that, as discussed in [9], a low-rate charge or discharge cannot be considered as OCV or pseudo-OCV curves as they are not independent of the regime. The pseudo-OCV curve can be calculated by averaging the low-rate charge and discharge curves [9,41,42]. This method is usually accurate, but it is highly recommended to check the validity of the results. First, the residual capacity measurements for the low-rate cycle must be small, <1% of the maximum capacity. High residual capacity measurement for the low-rate cycles would suggest that some SOC was not utilized during the low-rate cycles. If so, no data for this additional capacity were gathered and thus the corresponding OCV voltage cannot be assessed [31]. In such occurrence, it is recommended to lower the current of the low-rate cycle and of the residual capacity cycle. The second verification is to compare the SOC obtained by reporting the RCVs from both cycles on the pseudo-OCV curve with the SOC calculated from the residual capacity measurements divided by the maximum capacity. The values should be similar.

Using the SOC calculated at end of charge and end of discharge (either from the residual capacity measurements or the *pseudo*-OCV curve), the SOC windows used by the different rates can be compared and tracked [43] so that differences between rates, temperature, or SOH can be compared and discussed. An increasing difference between low and high rate implies growing kinetic limitations. A decreasing difference is possible in case of electrochemical milling enhancing the surface area of the electrode [44]. This allows the comparison of SOC (the percentage of the maximum lithiation) and the depth-of-discharge (DOD, the percentage of the maximum lithiation under a given duty cycle) [9]. For the data collected during the RPT protocol, each charge or discharge corresponds to 100% DOD since the cells were fully charged or discharged prior to the start of the next regime. More details on the extremely important question of SOC definition can be found in [9].

An example of the pseudo-OCV calculation and validation is presented in Figure 6 for a commercial graphite/nickel aluminum cobalt oxide cell. The residual capacities were below 0.5% of the maximum capacity, consequently the pseudo-OCV curve was calculated. This pseudo-OCV curve was then used to infer the SOC from the RCV. For both cycles, the estimated SOCs were within 0.5% from the residual capacity measurements (highlighted by the vertical bars). In that case, the pseudo-OCV curve could be deemed accurate enough for use. From the SOC windows, 100% DOD at C/25 corresponds to 99.5% SOC versus 95% SOC in discharge and 90% SOC in charge at C/3. Examples of results obtainable from these analyses can be found in our previous work including the SOC range variation with rate [43] and the RCV and OCV evolution upon aging [45], at different temperatures [46], and for different electrode architectures [44].

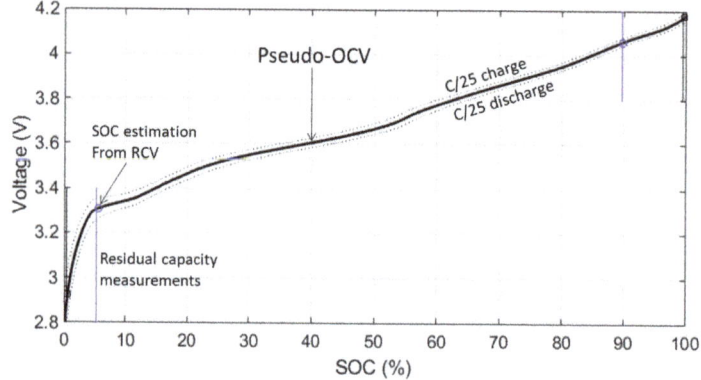

Figure 6. The C/25 charge and discharge curves and associated pseudo-open circuit voltage (OCV) curve. Black squares and blue circles indicate the state-of-charge (SOC) estimated from C/3 and C/25 rest cell voltages, respectively. Vertical lines indicate the amount of additional capacity added by the residual capacity steps.

The third set of information that can be gathered from the RPT is the voltage vs. capacity curves of the low and high-rate cycles. The low-rate cycle provides thermodynamic information; while the high-rate cycle reveals kinetic information. Since voltage variations are minute, derivatives are recommended, either incremental capacity (IC, $dQ/dV = f(V)$) [9,42,45,47–49] or differential voltage (DV, $dV/dQ = f(Q)$) [50–52]. Interested readers should refer to [9] for a complete discussion of the advantages and disadvantages of both these electrochemical voltage spectroscopies (EVS).

As mentioned in [9], there are two levels of analysis for the EVS curves. The qualitative way is the easiest and involves comparing the curves. Different features of interest (FOI) [53] (Figure 7a) can be discussed and characterized without a complete understanding of the underlying electrochemical process. To properly discuss changes, it is extremely important to describe the peaks properly (Figure 7b). By convention, for IC, discharge peaks are negative and charge peaks are positive. The peak potential is measured at the base of the peak and not at maximum intensity. Each peak corresponds to a thermodynamic property, a redox reaction, and thus the potential of the reaction is the same in charge and discharge. The voltage of the maximum of intensity could be affected by hysteresis [54], polarization, and kinetics [55]. For example, if a peak is broadening, the position of the intensity maximum changes, whereas the underlying reaction is still the same and thus starts at the same potential, Figure 7b. In that case, the slope of the front of the peak would be an interesting FOI to characterize kinetic changes. Another example is the shifting of all the peaks towards lower potential during discharge, which is usually a clear indication that the resistance is increasing (not observable on DV curves [9]). The resolution needed for IC and DV curves depends on the chemistry. Following common practices from X-ray diffraction studies, having 5 or 6 points above the half-width of the thinner peak would be considered sufficient resolution to trust peak intensity and position. Furthermore, 1 mV or 2 mV voltage steps usually provide good enough resolution for single cells, but this depends on chemistry and on the quality of the data. If noise filtering techniques are applied to clean the curves, particular attention must be set on possible peak displacements.

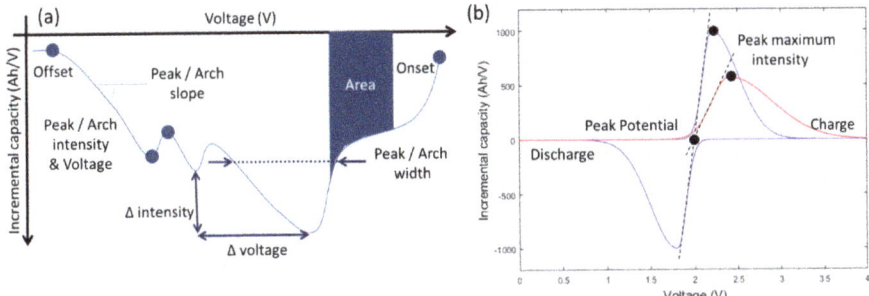

Figure 7. (**a**) Feature of interest (adapted from [53]) and (**b**) incremental capacity (IC) peak description.

In addition to the traditional IC and DV analysis, with the same dataset, other voltage derivative techniques can be useful in the study of relaxation curves and the temperature variations. Analysis of dV/dt = f(t) plots have shown the potential to identify lithium plating from relaxation curves [56]. Differential temperature curves with respect to voltage (dT/dV = f(V)) can also provide additional information on degradation mechanisms [57].

The quantitative analysis is much more complex, but it determines the magnitudes of the degradation modes such as loss of active material, loss of reactant, and kinetic degradation [55,58,59]. The methodology concerning the interpretation of the incremental capacity curves is out of the scope of this paper and was extensively described in previous publications [9,58,59] with the introduction of the clepsydra analogy that visualized the problem as communicating vessels with the liquid representing the lithium in the system and the shapes of the vessels defined by the derivative of the voltage responses for both electrodes (Figure 8a). This step can be bypassed by using one of the publicly available mechanistic models [59–61] that relies on electrode half-cell data to build a virtual replicate and emulate the impact of all degradation mechanisms based on simple parameters such as the loading ratio (LR) between the capacities of the positive and negative electrodes and their offset (OFS), Figure 8b. Every degradation induces changes to LR and OFS, Figure 8c, and those changes can be related to the degradation modes via a simple set of equations [59].

The typical way of performing the analysis of IC or DV curves is to first select some FOIs [53,62], then compare their experimental evolution to predicted ones for individual degradation modes, and finally validate by simulating the full degradation to match the entire voltage response of the cell. The full match might only be possible with the exact same positive and negative electrodes as the ones used in the considered cells. For this reason, we recommend harvesting electrodes from one cell and test them versus a reference electrode to later be used in the mechanistic model. Discussion on how to open commercial cells and perform half-cell testing is out of the scope of this paper and more details can be found in [29,63–69]. Individual electrodes in half-cell should be tested with an RPT protocol similar to the one presented in Figure 5 but with more rates ranging from twice as low to twice as fast to enable simulation of loss of active material and kinetic changes [55]. Since only one RPT is performed on the half-cells and the capacity at each rate is normalized from the RCVs, the residual capacity measurements and the OCV curve, degradation between cycles is not an issue. Thus, increasing the number of rates tested improves the accuracy of the simulations. However, some attention needs to be spent on the procedure to minimize cell-to-cell variations and improve the agreement of the data from the half-cells and the full cell [66]. With one RPT for the full cell and one RPT for each electrode, enough information is available to build a virtual cell and use the mechanistic modeling tools. If one of the electrodes is a blend of several active materials, the best results will be obtained if half-cell data are available for each individual electrode component. If the cells cannot be opened, or if individual components of blended electrodes cannot be gathered, the same analysis can be done using reference materials [70]. However, in that case, only trends can be used to compare the mechanistic simulations to the experimental data.

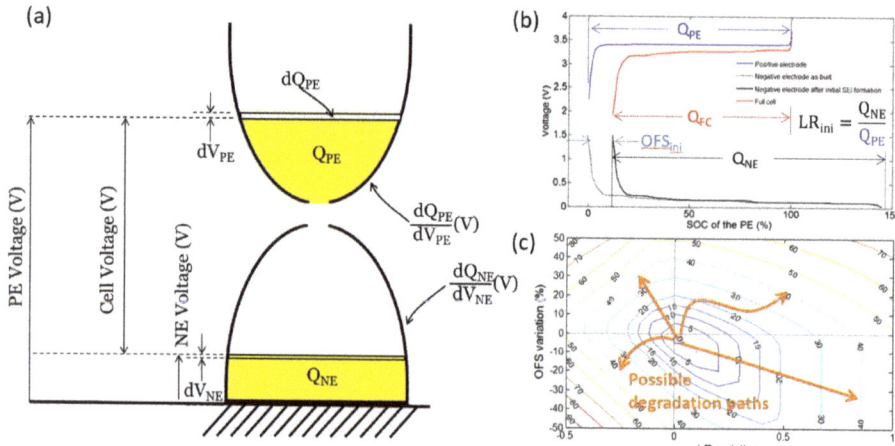

Figure 8. (a) Clepsydra analogy, (b) parameterization for mechanistic modeling, and (c) the effect of changes of loading ratio and electrode offset.

For analysis, the FOI selection must only be made after an exhaustive sensibility analysis [10,53]. It is highly recommended to first simulate a degradation map with the voltage changes and capacity loss associated with all the individual modes to gain an understanding of which FOI is more sensible to a degradation mode. Logical deductions from this degradation map usually allow the direct estimation of at least one of the degradation modes. With one quantified, another is then usually unambiguously decipherable and so on until only one is remaining. Quantifying the last degradation mode directly is usually not possible because of combined effects with the others. This quantification usually necessitates a full fit match. This could be achieved through an automated calculation. Some algorithms were proposed in the literature based on FOIs but most of them neglected the proper sensibility analysis and hence should not be trusted to be universal. A detailed discussion on the necessity for sensibility analysis is presented in [53]. Among the others, examples of degradation maps can be found in [71] for graphite//nickel cobalt aluminum oxide cells, in [53,59] for graphite//lithium iron phosphate cells, and in [53,72] for lithium titanate oxide//manganese nickel cobalt oxide. Example of studies with blended electrodes can be found in [70,73]. In our past IC studies [44,53,59,65,70–75], we often used different FOIs and entry points for the IC analysis depending on the chemistry and the experimental trends. This highlights that a step-by-step instruction list on how to perform the analysis is not possible and that it should always be guided by a sensibility analysis to be repeated for every new study. It must be noted that IC or DV analysis are usually chemistry specific and, as such, results should never be extrapolated to other chemistries without careful considerations and verifications.

2.4. Duty Cycle

The definition of the duty cycle test is the most important part of any degradation or validation study. Battery degradation is path dependent [10,74,76–80] meaning that different conditions are degrading battery differently, not just increasing or decreasing the rate. Therefore, to be representative of a given application, the duty cycle to be applied needs to be as close as possible to the real use, which is most likely stochastic [74,76–80]. Depending on the cells, some cycling profiles could be surprisingly harsh. For example, our fast charging studies showed no impact of 4C constant-current discharges compared to 1C but fast degradation under EV type pulsing discharges in which the average current was around C/3 and the maximum current 4C [10,65,75]. This stresses that significant time needs to be allocated to properly define the duty cycle based on the targeted application. If datasets are available, fuzzy logic [81] and statistical analysis [82] are good options to define a representative usage cycle. Alternatively, literature might suggest adapted duty cycles, although they might not be fully representative [74].

Even if a representative usage is determined, real-life conditions can fluctuate significantly. Therefore, it might be valuable to test conditions around the representative usage to look for optimal or detrimental conditions for performance or durability. Testing every possible condition is not possible. Fortunately, some statistical tools allow the sampling of a wide range of values for meaningful parameters with an optimal number of experiments. This is called design of experiments and it is effective in battery testing. Interested readers are referred to [83–86] for more details on how to set them up and how to interpret the results. Such a methodology was already proposed and successfully applied to battery testing in recent years [30,87–92].

The duty cycle is the center part of any study. The RPT will help diagnosis and enable prognosis on the cells but these results need to be relevant to different duty cycles. With a well-defined testing plan, some statistical analyses, such as the analysis of variance (ANOVA), can be executed to establish the significance of the different factors in the duty cycle (e.g., current, depth of discharge, and temperature). An example of significance analysis from [70] is presented in Figure 9. It shows the relative significance of the different factors: The SOC swings, the rate, and the temperature on the different degradation modes. Such results can then be used to derive the optimal conditions to limit capacity loss or the effectiveness of a given control algorithm.

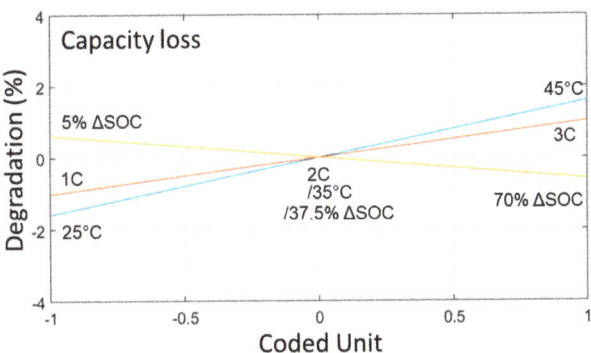

Figure 9. Example of a significance analysis taken from [70].

For studies focused on long-term ramifications, it is important to quantify the intrinsic variability in cycle aging and calendar aging, in other words, how much difference is observed between several cells performing the exact same duty cycle. Several studies reported a noticeable spread in cycle-lives [20,23,31,93,94] that could be detrimental to the durability of battery packs. Moreover, as discussed in [10], for the deployment of algorithms, validation on one duty cycle is not enough. The entirety of the degradation paths should be tested. If this is not possible experimentally, modeling solutions should be considered [10,95,96]. In some cases, the capacity is shown to be increased after performing an RPT compared to the last cycles before the RPT. This could be associated with negative electrode overhang [97,98] and its impact will usually fade away after a few cycles.

2.5. Post-Mortem

If necessary, and to validate the diagnosis gathered from the analysis of the RPT, some post-mortem tests can be carried out. This will require multiple apparatus and thus details are out of the scope of this publication. The range of post-mortem tests proposed in the literature is wide [11,12,99] and some techniques might require a lot of resources. They are typically performed by opening the aged cells to test individual electrodes but some can be performed on full cells [11]. Electrochemical tests on aged electrodes can verify changes in OFS and LR and validate the loss of reactant and loss of active material quantifications [64] and other tests can be used to discover the origin of the losses.

3. Conclusions

In summary, laboratory battery testing is a much more complex task than it appears. Properly defined and executed plans expedites a deeper understanding of the performance and SOH of commercial batteries. Inadequate validation can delegitimize a good study. This paper presents best practices for simple and effective testing of batteries. For the most part, execution only requires a multichannel potentiostat/galvanostat without the need for other complex instrumentation. This will allow characterization of not only the cell-to-cell variations, but also evolution of SOH throughout the lifetime of the cells. Although this publication was centered on single cell testing, the same approach can also be used for modules or packs if safety controls are in place so that no single cell can be overcharged or overdischarged.

Author Contributions: Conceptualization, M.D.; methodology, M.D.; software, M.D.; validation, M.D. and G.B.; formal analysis, M.D. and G.B.; resources, M.D.; data curation, M.D.; writing—original draft preparation, M.D.; writing—review and editing, M.D. and G.B.; visualization, M.D.; supervision, M.D.; project administration, M.D.; funding acquisition, M.D. All authors have read and agreed to the published version of the manuscript.

Funding: This work was funded by ONR Asia Pacific Research Initiative for Sustainable Energy Systems (APRISES), award number N00014-18-1-2127. M.D. is also supported by the State of Hawaii.

Acknowledgments: The authors are thankful to all the past staff that helped define and affine these protocols, most notably Bor Yann Liaw, Cyril Truchot, and Arnaud Devie. M.D. would also like to thank the University of Hawaii Material Science Consortium for Research and Education for advanced material characterizations.

Conflicts of Interest: The authors declare no conflict of interest.

References

1. Lee, T.; Glick, M.B.; Lee, J.-H. Island energy transition: Assessing Hawaii's multi-level, policy-driven approach. *Renew. Sustain. Energy Rev.* **2020**, *118*. [CrossRef]
2. Zhao, W.; Yi, J.; He, P.; Zhou, H. Solid-state electrolytes for lithium-ion batteries: Fundamentals, challenges and perspectives. *Electrochem. Energy Rev.* **2019**, *2*, 574–605. [CrossRef]
3. Wang, L.; Wu, Z.; Zou, J.; Gao, P.; Niu, X.; Li, H.; Chen, L. Li-free cathode materials for high energy density lithium batteries. *Joule* **2019**, *3*, 2086–2102. [CrossRef]
4. Shen, M.; Gao, Q. A review on battery management system from the modeling efforts to its multiapplication and integration. *Int. J. Energy Res.* **2019**. [CrossRef]
5. Plett, G.L. Review and some perspectives on different methods to estimate state of charge of lithium-Ion batteries. *J. Automot. Saf. Energy* **2019**, *10*, 249–272. [CrossRef]
6. Meng, H.; Li, Y.-F. A review on prognostics and health management (PHM) methods of lithium-ion batteries. *Renew. Sustain. Energy Rev.* **2019**, *116*. [CrossRef]
7. Lin, Q.; Wang, J.; Xiong, R.; Shen, W.; He, H. Towards a smarter battery management system: A critical review on optimal charging methods of lithium ion batteries. *Energy* **2019**. [CrossRef]
8. Li, Y.; Liu, K.; Foley, A.M.; Zülke, A.; Berecibar, M.; Nanini-Maury, E.; Van Mierlo, J.; Hoster, H.E. Data-driven health estimation and lifetime prediction of lithium-ion batteries: A review. *Renew. Sustain. Energy Rev.* **2019**, *113*. [CrossRef]
9. Barai, A.; Uddin, K.; Dubarry, M.; Somerville, L.; McGordon, A.; Jennings, P.; Bloom, I. A comparison of methodologies for the non-invasive characterisation of commercial Li-ion cells. *Progr. Energy Combust. Sci.* **2019**, *72*, 1–31. [CrossRef]
10. Dubarry, M.; Baure, G.; Anseán, D. Perspective on state of health determination in lithium ion batteries. *J. Electrochem. Energy Convers. Storage* **2020**, 1–25, in press. [CrossRef]

11. Waldmann, T.; Iturrondobeitia, A.; Kasper, M.; Ghanbari, N.; Aguesse, F.; Bekaert, E.; Daniel, L.; Genies, S.; Jimenez Gordon, I.; Loble, M.; et al. Review—Post-mortem analysis of aged lithium-ion batteries: Disassembly methodology and physico-chemical analysis techniques. *J. Electrochem. Soc.* **2016**, *163*, A2149–A2164. [CrossRef]
12. Lu, J.; Wu, T.; Amine, K. State-of-the-art characterization techniques for advanced lithium-ion batteries. *Nat. Energy* **2017**, *2*, 17011. [CrossRef]
13. Taylor, J.; Barai, A.; Ashwin, T.R.; Guo, Y.; Amor-Segan, M.; Marco, J. An insight into the errors and uncertainty of the lithium-ion battery characterisation experiments. *J. Energy Storage* **2019**, *24*. [CrossRef]
14. De-Leon, S. Battery safety training for portable & stationary applications. In Proceedings of the Next Generation Energy Storage, San Diego, CA, USA, 18–20 April 2016.
15. Dubarry, M.; Qin, N.; Brooker, P. Calendar aging of commercial Li-ion cells of different chemistries—A review. *Curr. Opin. Electrochem.* **2018**, *9*, 106–113. [CrossRef]
16. Feng, X.; Ouyang, M.; Liu, X.; Lu, L.; Xia, Y.; He, X. Thermal runaway mechanism of lithium ion battery for electric vehicles: A review. *Energy Storage Mater.* **2017**. [CrossRef]
17. Börger, A.; Mertens, J.; Wenzl, H. Thermal runaway and thermal runaway propagation in batteries: What do we talk about? *J. Energy Storage* **2019**, *24*. [CrossRef]
18. Wang, Q.; Mao, B.; Stoliarov, S.I.; Sun, J. A review of lithium ion battery failure mechanisms and fire prevention strategies. *Progr. Energy Combust. Sci.* **2019**, *73*, 95–131. [CrossRef]
19. Wu, X.; Song, K.; Zhang, X.; Hu, N.; Li, L.; Li, W.; Zhang, L.; Zhang, H. Safety issues in lithium Ion batteries: Materials and cell design. *Front. Energy Res.* **2019**, *7*. [CrossRef]
20. Cripps, E.; Pecht, M. A bayesian nonlinear random effects model for identification of defective batteries from lot samples. *J. Power Sources* **2017**, *342*, 342–350. [CrossRef]
21. An, F.; Chen, L.; Huang, J.; Zhang, J.; Li, P. Rate dependence of cell-to-cell variations of lithium-ion cells. *Sci. Rep.* **2016**, *6*, 35051. [CrossRef]
22. Schuster, S.F.; Brand, M.J.; Berg, P.; Gleissenberger, M.; Jossen, A. Lithium-ion cell-to-cell variation during battery electric vehicle operation. *J. Power Sources* **2015**, *297*, 242–251. [CrossRef]
23. Baumhöfer, T.; Brühl, M.; Rothgang, S.; Sauer, D.U. Production caused variation in capacity aging trend and correlation to initial cell performance. *J. Power Sources* **2014**, *247*, 332–338. [CrossRef]
24. Santhanagopalan, S.; White, R.E. Quantifying cell-to-cell variations in lithium Ion batteries. *Int. J. Electrochem.* **2012**, *2012*, 1–10. [CrossRef]
25. Kim, J.; Shin, J. Screening process of Li-ion series battery pack for improved voltage soc balancing. In Proceedings of the International Power Electronics Conference, Sapporo, Japan, 21–24 June 2010.
26. Rumpf, K.; Naumann, M.; Jossen, A. Experimental investigation of parametric cell-to-cell variation and correlation based on 1100 commercial lithium-ion cells. *J. Energy Storage* **2017**, *14*, 224–243. [CrossRef]
27. Robertson, D.C.; Christophersen, J.P.; Bennett, T.; Walker, L.K.; Wang, F.; Liu, S.; Fan, B.; Bloom, I. A comparison of battery testing protocols: Those used by the U.S. advanced battery consortium and those used in China. *J. Power Sources* **2016**, *306*, 268–273. [CrossRef]
28. Dubarry, M.; Vuillaume, N.; Liaw, B.Y. Origins and accommodation of cell variations in Li-ion battery pack modeling. *Int. J. Energy Res.* **2010**, *34*, 216–231. [CrossRef]
29. Devie, A.; Dubarry, M. Durability and reliability of electric vehicle batteries under electric utility grid operations. Part 1: Cell-to-cell variations and preliminary testing. *Batteries* **2016**, *2*, 28. [CrossRef]
30. Dubarry, M.; Devie, A. Battery durability and reliability under electric utility grid operations: Representative usage aging and calendar aging. *J. Energy Storage* **2018**, *18*, 185–195. [CrossRef]
31. Devie, A.; Baure, G.; Dubarry, M. Intrinsic variability in the degradation of a batch of commercial 18650 Lithium-Ion cells. *Energies* **2018**, *11*, 1031. [CrossRef]
32. Wood, D.L.; Li, J.; An, S.J. Formation challenges of lithium-ion battery manufacturing. *Joule* **2019**, *3*, 2884–2888. [CrossRef]
33. Heubner, C.; Schneider, M.; Michaelis, A. Diffusion-limited c-rate: A fundamental principle quantifying the intrinsic limits of Li-Ion batteries. *Adv. Energy Mater.* **2019**. [CrossRef]
34. Peukert, W. An equation forrelating capacity to discharge rate. *Electrotech. Z.* **1897**, *1*, 287–288.
35. Bard, A.; Faulkner, L. *Electrochemical Methods—Fundamentals and Applications*, 2nd ed.; Wiley: Hoboken, NJ, USA, 2001.

36. Liaw, B.Y.; Dubarry, M. A roadmap to understand battery performance in electric and hybrid vehicle operation. In *Electric and Hybrid Vehicles*; Pistoia, G., Ed.; Elsevier: Amsterdam, The Netherlands, 2010; pp. 375–403. [CrossRef]
37. Christophersen, J.P.; Ho, C.D.; Motloch, C.G.; Howell, D.; Hess, H.L. Effects of reference performance testing during aging using commercial Lithium-Ion cells. *J. Electrochem. Soc.* **2006**, *153*, A1406. [CrossRef]
38. INL. *Battery Test Manual For Electric Vehicles*; INL: Idaho Falls, ID, USA, 2015.
39. Soto, A.; Berrueta, A.; Sanchis, P.; Ursúa, A. Analysis of the main battery characterization techniques and experimental comparison of commercial 18650 Li-ion cells. In Proceedings of the 2019 IEEE International Conference on Environment and Electrical Engineering and 2019 IEEE Industrial and Commercial Power Systems Europe (EEEIC/I&CPS Europe), Genova, Italy, 11–14 June 2019.
40. Liu, C.; Neale, Z.G.; Cao, G. Understanding electrochemical potentials of cathode materials in rechargeable batteries. *Mater. Today* **2016**, *19*, 109–123. [CrossRef]
41. Truchot, C.; Dubarry, M.; Liaw, B.Y. State-of-charge estimation and uncertainty for lithium-ion battery strings. *Appl. Energy* **2014**, *119*, 218–227. [CrossRef]
42. Dubarry, M.; Truchot, C.; Cugnet, M.; Liaw, B.Y.; Gering, K.; Sazhin, S.; Jamison, D.; Michelbacher, C. Evaluation of commercial lithium-ion cells based on composite positive electrode for plug-in hybrid electric vehicle applications. Part I: Initial characterizations. *J. Power Sources* **2011**, *196*, 10328–10335. [CrossRef]
43. Dubarry, M.; Svoboda, V.; Hwu, R.; Liaw, B.Y. Capacity loss in rechargeable lithium cells during cycle life testing: The importance of determining state-of-charge. *J. Power Sources* **2007**, *174*, 1121–1125. [CrossRef]
44. Dubarry, M.; Truchot, C.; Liaw, B.Y. Cell degradation in commercial LiFePO4 cells with high-power and high-energy designs. *J. Power Sources* **2014**, *258*, 408–419. [CrossRef]
45. Dubarry, M.; Truchot, C.; Liaw, B.Y.; Gering, K.; Sazhin, S.; Jamison, D.; Michelbacher, C. Evaluation of commercial lithium-ion cells based on composite positive electrode for plug-in hybrid electric vehicle applications. Part II. Degradation mechanism under 2C cycle aging. *J. Power Sources* **2011**, *196*, 10336–10343. [CrossRef]
46. Dubarry, M.; Truchot, C.; Liaw, B.Y.; Gering, K.; Sazhin, S.; Jamison, D.; Michelbacher, C. Evaluation of commercial lithium-ion cells based on composite positive electrode for plug-in hybrid electric vehicle applications: III. Effect of thermal excursions without prolonged thermal aging. *J. Electrochem. Soc.* **2013**, *160*, A191–A199. [CrossRef]
47. Balewski, L.; Brenet, J.P. A new method for the study of the electrochemical reactivity of manganese dioxide. *Electrochem. Technol.* **1967**, *5*, 527–531.
48. Dubarry, M.; Svoboda, V.; Hwu, R.; Liaw, B.Y. Incremental capacity analysis and close-to-equilibrium OCV measurements to quantify capacity fade in commercial rechargeable lithium batteries. *Electrochem. Solid State Lett.* **2006**, *9*, A454–A457. [CrossRef]
49. Dubarry, M.; Liaw, B.Y. Identify capacity fading mechanism in a commercial LiFePO4 cell. *J. Power Sources* **2009**, *194*, 541–549. [CrossRef]
50. Bloom, I.; Christophersen, J.; Gering, K. Differential voltage analyses of high-power, lithium-ion cells. 2. Applications. *J. Power Sources* **2005**, *139*, 304–313. [CrossRef]
51. Bloom, I.; Jansen, A.N.; Abraham, D.P.; Knuth, J.; Jones, S.A.; Battaglia, V.S.; Henriksen, G.L. Differential voltage analyses of high-power, lithium-ion cells. 1. Technique and applications. *J. Power Sources* **2005**, *139*, 295–303. [CrossRef]
52. Bloom, I.; Christophersen, J.P.; Abraham, D.P.; Gering, K.L. Differential voltage analyses of high-power, lithium-ion cells. 3. Another anode phenomenon. *J. Power Sources* **2006**, *157*, 537–542. [CrossRef]
53. Dubarry, M.; Berecibar, M.; Devie, A.; Anseán, D.; Omar, N.; Villarreal, I. State of health battery estimator enabling degradation diagnosis: Model and algorithm description. *J. Power Sources* **2017**, *360*, 59–69. [CrossRef]
54. Dubarry, M.; Gaubicher, J.; Guyomard, D.; Wallez, G.; Quarton, M.; Baehtz, C. Uncommon potential hysteresis in the Li/Li2xVO(H2−xPO4)2 (0 ≤ x ≤ 2) system. *Electrochim. Acta* **2008**, *53*, 4564–4572. [CrossRef]
55. Schindler, S.; Baure, G.; Danzer, M.A.; Dubarry, M. Kinetics accommodation in Li-ion mechanistic modeling. *J. Power Sources* **2019**, *440*, 227117. [CrossRef]
56. Schindler, S.; Bauer, M.; Petzl, M.; Danzer, M.A. Voltage relaxation and impedance spectroscopy as in-operando methods for the detection of lithium plating on graphitic anodes in commercial lithium-ion cells. *J. Power Sources* **2016**, *304*, 170–180. [CrossRef]

57. Wu, B.; Yufit, V.; Merla, Y.; Martinez-Botas, R.F.; Brandon, N.P.; Offer, G.J. Differential thermal voltammetry for tracking of degradation in lithium-ion batteries. *J. Power Sources* **2015**, *273*, 495–501. [CrossRef]
58. Dubarry, M.; Devie, A.; Liaw, B.Y. The value of battery diagnostics and prognostics. *J. Energy Power Sources* **2014**, *1*, 242–249.
59. Dubarry, M.; Truchot, C.; Liaw, B.Y. Synthesize battery degradation modes via a diagnostic and prognostic model. *J. Power Sources* **2012**, *219*, 204–216. [CrossRef]
60. Dahn, H.M.; Smith, A.J.; Burns, J.C.; Stevens, D.A.; Dahn, J.R. User-friendly differential voltage analysis freeware for the analysis of degradation mechanisms in Li-Ion batteries. *J. Electrochem. Soc.* **2012**, *159*, A1405–A1409. [CrossRef]
61. HNEI. Alawa Central. Available online: https://www.soest.hawaii.edu/HNEI/alawa/ (accessed on 9 January 2020).
62. Berecibar, M.; Devriendt, F.; Dubarry, M.; Villarreal, I.; Omar, N.; Verbeke, W.; Van Mierlo, J. Online state of health estimation on NMC cells based on predictive analytics. *J. Power Sources* **2016**, *320*, 239–250. [CrossRef]
63. Abraham, D.P.; Knuth, J.L.; Dees, D.W.; Bloom, I.; Christophersen, J.P. Performance degradation of high-power lithium-ion cells—Electrochemistry of harvested electrodes. *J. Power Sources* **2007**, *170*, 465–475. [CrossRef]
64. Kassem, M.; Delacourt, C. Postmortem analysis of calendar-aged graphite/LiFePO4 cells. *J. Power Sources* **2013**, *235*, 159–171. [CrossRef]
65. Anseán, D.; Dubarry, M.; Devie, A.; Liaw, B.Y.; García, V.M.; Viera, J.C.; González, M. Fast charging technique for high power LiFePO4 batteries: A mechanistic analysis of aging. *J. Power Sources* **2016**, *321*, 201–209. [CrossRef]
66. Schmid, A.U.; Kurka, M.; Birke, K.P. Reproducibility of Li-ion cell reassembling processes and their influence on coin cell aging. *J. Energy Storage* **2019**, *24*. [CrossRef]
67. Murray, V.; Hall, D.S.; Dahn, J.R. A guide to full coin cell making for academic researchers. *J. Electrochem. Soc.* **2019**, *166*, A329–A333. [CrossRef]
68. Zhou, G.; Wang, Q.; Wang, S.; Ling, S.; Zheng, J.; Yu, X.; Li, H. A facile electrode preparation method for accurate electrochemical measurements of double-side-coated electrode from commercial Li-ion batteries. *J. Power Sources* **2018**, *384*, 172–177. [CrossRef]
69. Wu, B.; Yang, Y.; Liu, D.; Niu, C.; Gross, M.; Seymour, L.; Lee, H.; Le, P.M.L.; Vo, T.D.; Deng, Z.D.; et al. Good practices for rechargeable lithium metal batteries. *J. Electrochem. Soc.* **2019**, *166*, A4141–A4149. [CrossRef]
70. Baure, G.; Devie, A.; Dubarry, M. Battery durability and reliability under electric utility grid operations: Path dependence of battery degradation. *J. Electrochem. Soc.* **2019**, *166*, A1991–A2001. [CrossRef]
71. Dubarry, M.; Baure, G.; Devie, A. Durability and reliability of EV batteries under electric utility grid operations: Path dependence of battery degradation. *J. Electrochem. Soc.* **2018**, *165*, A773–A783. [CrossRef]
72. Devie, A.; Dubarry, M.; Liaw, B.Y. Overcharge study in Li4Ti5O12 based Lithium-Ion pouch cell: I. Quantitative diagnosis of degradation modes. *J. Electrochem. Soc.* **2015**, *162*, A1033–A1040. [CrossRef]
73. Anseán, D.; Baure, G.; González, M.; Cameán, I.; García, A.B.; Dubarry, M. Mechanistic investigation of Silicon–Graphite//LiNi0.8Mn0.1Co0.1O2 commercial cells for non-intrusive diagnosis and prognosis. *J. Power Sources* **2020**. submitted.
74. Baure, G.; Dubarry, M. Synthetic vs. real driving cycles: A comparison of electric vehicle battery degradation. *Batteries* **2019**, *5*, 42. [CrossRef]
75. Anseán, D.; Dubarry, M.; Devie, A.; Liaw, B.Y.; García, V.M.; Viera, J.C.; González, M. Operando lithium plating quantification and early detection of a commercial LiFePO 4 cell cycled under dynamic driving schedule. *J. Power Sources* **2017**, *356*, 36–46. [CrossRef]
76. Gering, K.L.; Sazhin, S.V.; Jamison, D.K.; Michelbacher, C.J.; Liaw, B.Y.; Dubarry, M.; Cugnet, M. Investigation of path dependence in commercial lithium-ion cells chosen for plug-in hybrid vehicle duty cycle protocols. *J. Power Sources* **2011**, *196*, 3395–3403. [CrossRef]
77. Wu, W.; Wu, W.; Qiu, X.; Wang, S. Low-temperature reversible capacity loss and aging mechanism in lithium-ion batteries for different discharge profiles. *Int. J. Energy Res.* **2018**, *43*, 243–253. [CrossRef]
78. Radhakrishnan, K.N.; Coupar, T.; Nelson, D.J.; Ellis, M.W. Experimental evaluation of the effect of cycle profile on the durability of commercial Lithium Ion power cells. *J. Electrochem. Energy Convers. Storage* **2019**, *16*. [CrossRef]
79. Klett, M.; Eriksson, R.; Groot, J.; Svens, P.; Ciosek Högström, K.; Lindström, R.W.; Berg, H.; Gustafson, T.; Lindbergh, G.; Edström, K. Non-uniform aging of cycled commercial LiFePO4//graphite cylindrical cells revealed by post-mortem analysis. *J. Power Sources* **2014**, *257*, 126–137. [CrossRef]

80. Keil, P.; Jossen, A. Charging protocols for lithium-ion batteries and their impact on cycle life—An experimental study with different 18650 high-power cells. *J. Energy Storage* **2016**, *6*, 125–141. [CrossRef]
81. Liaw, B.Y.; Dubarry, M. From driving cycle analysis to understanding battery performance in real-life electric hybrid vehicle operation. *J. Power Sources* **2007**, *174*, 76–88. [CrossRef]
82. Dubarry, M.; Devie, A.; Stein, K.; Tun, M.; Matsuura, M.; Rocheleau, R. Battery energy storage system battery durability and reliability under electric utility grid operations: Analysis of 3 years of real usage. *J. Power Sources* **2017**, *338*, 65–73. [CrossRef]
83. Montgomery, D. *Design and Analysis of Experiments*, 8th ed.; Wiley: Hoboken, NJ, USA, 2013.
84. Antony, J. *Design of Experiments for Engineers and Scientists*; Elsevier Science & Technology Books: Amsterdam, The Netherlands, 2003.
85. Rynne, O.; Dubarry, M.; Molson, C.; Nicolas, E.; Lepage, D.; Prébé, A.; Aymé-Perrot, D.; Rochefort, D.; Dollé, M. Designs of experiments to optimize Li-ion battery electrodes' formulation. *J. Electrochem. Soc.* **2020**, submitted.
86. Rynne, O.; Dubarry, M.; Molson, C.; Lepage, D.; Prébé, A.; Aymé-Perrot, D.; Rochefort, D.; Dollé, M. Designs of experiments for beginners—A quick start guide for application to electrode formulation. *Batteries* **2019**, *5*, 72. [CrossRef]
87. Su, L.; Zhang, J.; Wang, C.; Zhang, Y.; Li, Z.; Song, Y.; Jin, T.; Ma, Z. Identifying main factors of capacity fading in lithium ion cells using orthogonal design of experiments. *Appl. Energy* **2016**, *163*, 201–210. [CrossRef]
88. Cui, Y.; Du, C.; Yin, G.; Gao, Y.; Zhang, L.; Guan, T.; Yang, L.; Wang, F. Multi-stress factor model for cycle lifetime prediction of lithium ion batteries with shallow-depth discharge. *J. Power Sources* **2015**, *279*, 123–132. [CrossRef]
89. Prochazka, W.; Pregartner, G.; Cifrain, M. Design-of-experiment and statistical modeling of a large scale aging experiment for two popular Lithium Ion cell chemistries. *J. Electrochem. Soc.* **2013**, *160*, A1039–A1051. [CrossRef]
90. Dubarry, M.; Devie, A.; McKenzie, K. Durability and reliability of electric vehicle batteries under electric utility grid operations: Bidirectional charging impact analysis. *J. Power Sources* **2017**, *358*, 39–49. [CrossRef]
91. Mathieu, R.; Baghdadi, I.; Briat, O.; Gyan, P.; Vinassa, J.-M. D-optimal design of experiments applied to lithium battery for ageing model calibration. *Energy* **2017**, *141*, 2108. [CrossRef]
92. Baghdadi, I.; Mathieu, R.; Briat, O.; Gyan, P.; Vinassa, J.-M. Lithium-ion battery ageing assessment based on a reduced design of experiments. In Proceedings of the 2017 IEEE Vehicle Power and Propulsion Conference (VPPC), Belfort, France, 11–14 December 2017.
93. Rohr, S.; Müller, S.; Baumann, M.; Kerler, M.; Ebert, F.; Kaden, D.; Lienkamp, M. Quantifying uncertainties in reusing Lithium-Ion batteries from electric vehicles. *Procedia Manuf.* **2017**, *8*, 603–610. [CrossRef]
94. Harris, S.J.; Harris, D.J.; Li, C. Failure statistics for commercial lithium ion batteries: A study of 24 pouch cells. *J. Power Sources* **2017**, *342*, 589–597. [CrossRef]
95. Dubarry, M.; Pastor-Fernández, C.; Baure, G.; Yu, T.F.; Widanage, W.D.; Marco, J. Battery energy storage system modeling: Investigation of intrinsic cell-to-cell variations. *J. Energy Storage* **2019**, *23*, 19–28. [CrossRef]
96. Dubarry, M.; Baure, G.; Pastor-Fernández, C.; Yu, T.F.; Widanage, W.D.; Marco, J. Battery energy storage system modeling: A combined comprehensive approach. *J. Energy Storage* **2019**, *21*, 172–185. [CrossRef]
97. Lewerenz, M.; Fuchs, G.; Becker, L.; Sauer, D.U. Irreversible calendar aging and quantification of the reversible capacity loss caused by anode overhang. *J. Energy Storage* **2018**, *18*, 149–159. [CrossRef]
98. Lewerenz, M.; Warnecke, A.; Sauer, D.U. Introduction of capacity difference analysis (CDA) for analyzing lateral Lithium-Ion flow to determine the state of covering layer evolution. *J. Power Sources* **2017**, *354*, 157–166. [CrossRef]
99. Kovachev, G.; Schröttner, H.; Gstrein, G.; Aiello, L.; Hanzu, I.; Wilkening, H.M.R.; Foitzik, A.; Wellm, M.; Sinz, W.; Ellersdorfer, C. Analytical dissection of an automotive Li-Ion pouch cell. *Batteries* **2019**, *5*, 67. [CrossRef]

© 2020 by the authors. Licensee MDPI, Basel, Switzerland. This article is an open access article distributed under the terms and conditions of the Creative Commons Attribution (CC BY) license (http://creativecommons.org/licenses/by/4.0/).

Article

Battery Charging Procedure Proposal Including Regeneration of Short-Circuited and Deeply Discharged LiFePO₄ Traction Batteries

Pavol Spanik, Michal Frivaldsky *, Juraj Adamec and Matus Danko

Department of Mechatronics and Electronics, Faculty of Electrical Engineering and Information Technologies, University of Zilina, 010 26 Zilina, Slovakia; pavol.spanik@feit.uniza.sk (P.S.); juraj.adamec@feit.uniza.sk (J.A.); matus.danko@feit.uniza.sk (M.D.)
* Correspondence: michal.frivaldsky@fel.uniza.sk

Received: 20 April 2020; Accepted: 29 May 2020; Published: 2 June 2020

Abstract: The presented paper discusses the most often damages applying for lithium traction and non-traction cells. The focus is therefore given on investigation of possibilities related to the recovery of such damaged lithium-ion batteries, more specifically after long-term short-circuit and deep discharge. For this purpose, initially, the short-circuit was applied to the selected type of traction LiFePO₄ cell. Also, the deeply discharged cell was identified and observed. Both damaged cells would exhibit visible damage if electro-mechanical properties were measured. Individual types of damage require a different approach for battery regeneration to recover cells as much as possible. For this purpose, experimental set-up for automated system integrating proposed recovery methods were realized, while battery under test undergone a full-range of regeneration procedure. As a verification of the proposed regeneration algorithms, the test of delivered Ampere-hours (Ah) for various discharging currents was realized both for short-circuited as well as deeply discharged cells. Received results have been compared to the new/referenced cell, which undergoes the same test of delivered Ah. From the final evaluation is seen, that proposed procedure can recover damaged cell up to 80% of its full capacity if short-circuit was applied, or 70% if a deeply discharged cell is considered.

Keywords: traction battery; LiFePO₄; short-circuit; deep discharge; damage recovery

1. Introduction

Li-ion batteries have successfully penetrated our daily lives, from 3C products to EVs. For example, the newly developed Lithium Iron Phosphate (LiFePO₄) battery of larger energy capacity and safer chemical characteristics has been considered as an excellent power resource for future EVs. Anxiously, the capital costs of those Li-ion batteries are not low enough yet to substantially decrease the total EV cost for attracting numerous customers. Consequently, this fatal bottleneck directly impacts the universal adaptability of EVs. Some investigations delivered that the Li-ion battery cost needs to be chopped down around 50% for coming to EVs to compete against conventionally fueled vehicles with great odds [1–3] fully. Nowadays, in Europe, around 200 million vehicles are registered (approximately 80% are cars and 20% commercial vehicles). Car batteries deteriorate after two to three years, and people throw them away [4]. If only one battery from a total of five is replaced within a year, 15 million batteries are collected per year. Nevertheless, more than 15 million batteries are discarded only in Europe. Unfortunately, only one-third of the total quantity is recycled by manufacturers, and the rest is either eliminated or dumped in forests, rivers, and other places [5–7]. A significant reason for this increase in the number of used batteries is the short life cycle of lead-acid batteries. Besides, large-size batteries used for electric trucks are costly, with costs varying between 300 and 1500 euros, and hence involve very high periodic expenses and then, more expensive products. Preventing the damage of this battery type and extending its life cycle will have a high economic impact [8–10].

Recycling, as well as regeneration/recovery of the used or harmed battery cells is, therefore, a topic, which must be accepted if sustainability related to environment and costs is considered [11–13]. Currently, the regeneration processes are available mostly for the lead-acid and NiMH batteries used for large electricity storage systems or hybrid vehicles. Target devices are batteries when both electrode plates are settled by crystallizes and electrolyte is composed solely of distilled water. A traditional battery charger will not be able to charge such a battery because it requires regeneration. The regeneration process consists commonly of a set of high-powered electrical pulses that are breaking down the crystallized layers. Battery regenerator manufacturers due to company know-how do not describe regeneration procedures, and each applies a separately developed system. Based on this, it is difficult to observe and determine the exact methodology used for regeneration [14–16].

Regarding the perspective of lithium-based cells used for e-mobility, the topic of regeneration will come to the forefront as the price of lithium is still high. Consequently, the manufacturing process of traction batteries represents an expensive procedure. Nowadays, many analyses, model development, and estimation algorithms for the state of charge and state of health have been developed, enabling to precisely estimate the operational life of traction energy storage systems [17–19]. However, there is a lack of studies dealing with the possibilities of regeneration of damaged cells, which after the correct capacity recovery process, could be used secondarily in the energy storage process.

Therefore, within the presented paper, the experimental methodology for damaged battery (focus is given on LiFePO$_4$ traction cells) considering various types of damage are introduced. Initially, the test with overcharging is provided, in order to verify protection components of investigated cells, and to verify the fact, whether regeneration on the overcharged battery can be applied. Consequently, the investigation of long-term short-circuit is presented together with the proposal and evaluation of the recovery algorithm suited for this type of battery damage. The proposed procedure is derived from the principles of regeneration of lead-acid batteries. In contrast, individual methods (regeneration from short-circuiting and regeneration from deep discharge) have been tested experimentally through variation of amplitudes of pulses, duration of pulses, and repetition (frequency) of pulses.

Similarly, the recovery procedure is proposed on the cells that take inappropriate long-term storage, where the deep discharge of the cell can apply. This situation is also described, while the recovery procedure with settings relevant for a deep discharge state was applied to damaged cells and consequently evaluated through the test of delivered ampere-hours. Received results have been compared to new cells and also evaluated after continual use (30 days of charging and discharging under various load) of regenerated pieces.

2. Electrical Types of Damage of Traction Batteries

Regarding the operation of batteries, there are several hazardous conditions related to the electrical behavior of the circuit, which can cause damage to the battery itself. Consequences coming from the wrong operation primarily reflect into the loss of the capacity or expressive, open-circuit voltage (OCV) drop. If the excessive duration of the hazardous operation is lasting, it can cause secondary harm to the internal structure as well as the mechanical cover of the battery. Here it is discussed long-term short-circuit operation, overcharging of the battery above permissible voltage level, or excessive deep discharge during battery operation or/and battery improper storage. Because each of the mentioned unwanted operational conditions reflects in battery damage, it is valuable to find whether the suitable procedure can be applied for battery regeneration back to its nominal operational state. In this article, the attention is focused on the investigation of the impact rate of improper operation, i.e., overcharging and short-circuit of 40 Ah–128 Wh Sinopoly LiFePO$_4$ 3.2 V battery. At the same time, a regeneration algorithm is applied on long term short-circuited cell. Similarly, deeply discharged cell WINA 60 Ah LiFePO$_4$ 3.2 V was observed and consequently subjected to the application of the proposed regeneration procedure in order to restore its capacity and functionality. The main electrical parameters of both tested cells are listed in Table 1.

Table 1. Technical parameters of investigated cell Sinopoly LiFePO$_4$ 3.2 V 40 Ah.

Electrical Parameter	Sinopoly LiFePO$_4$ 3.2 V 40 Ah	WINALiFePO$_4$ 3.2 V 60 Ah	
Nominal voltage	3.2	3.2	(V)
Maximum charging voltage	4	3.8	(V)
Minimum voltage	2.5	2.5	(V)
Maximum discharge current(continuous)	3	3	(C)
Optimal discharge current	13	20	(A)
Maximum charging current	80	90	(A)
Optimal charging current	13	20	(A)
Operating temperature	−45 to +85	−20 to +50	(°C)
Capacity	40	60	(Ah)
Shell material (package)	plastic	aluminum	(-)

2.1. Experimental Application of Overcharging

The experimental investigation of the first improper operational condition, i.e., overcharging of the battery, was realized with the use of test-stand, which is principally shown in Figure 1. The main device responsible for the simulation of the unwanted conditions is represented by programable power supply EA PSI 8080 (EA Elektro-Automatik GmbH & Co.KG, Viersen, Germany). It provides possibilities of programming of its output variables through instruction file, which covers information related to maximal values of charging voltage, current, and power as well for the predefined time interval. Recorded data of individual variables are stored in PC. For the analysis of visual damage observed over time, Canon EOS 6D is capturing images every 5 s, while thermal cameras FLIR E5 (FLIR, Boston, MA, USA). and FLIR SC660 (FLIR, Boston, MA, USA). are serving for detailed analysis of thermal performance during this experiment.

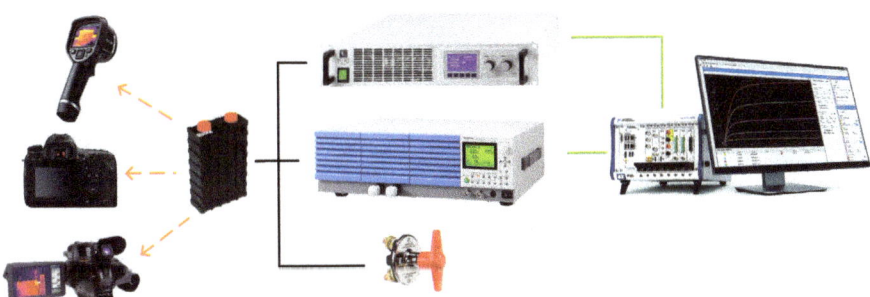

Figure 1. Block diagram of experimental test set-up for selected battery testing.

At the beginning of the test, the OCV of the battery was 3.43 V; thus, it refers to the charged state. According to the datasheet of the cell, the value of the charging voltage must not exceed 3.65 V. Initially, the selected overcharge value was set to 4.25 V, and the charging current was 10 A. CC&CV (Constant Current and Constant Voltage) charging mode was applied (Figure 2).

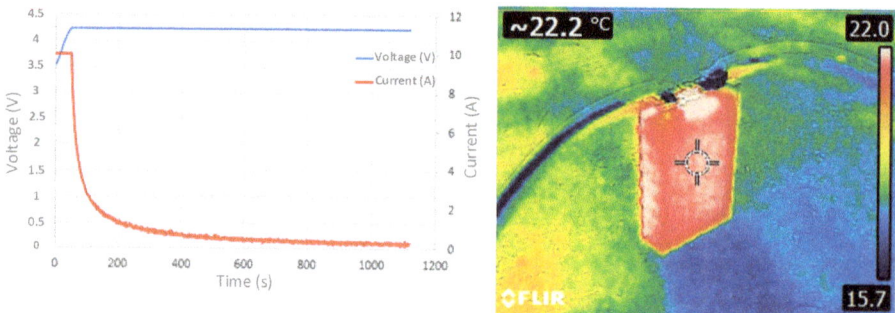

Figure 2. Waveforms of charging current and voltage during overcharging concerning for 4.25 V of charging voltage level.

The voltage on the battery reached the value of 4.25 V within the 50 s at a constant current of 10 A. From this point, the battery voltage was 4.25 V while the current was dropping gradually to almost 0 A. During the test, the temperature of the cell was maintained within the safe operating interval. At the same time, the maximum of 24.2 °C was achieved after 18 min and 36 s what represents 2.3 °C compared to the start of the test. During the experiment, no visible damage to the battery package was observed, and at the same time, no activation of the safety valve occurs. This test has continued with an increased level of charging voltage and current utilizing the same cell. Initially, for 35 s, 6.25 V/10 A was applied, while after 35 s, the level of charging current was increased to 30 A (Figure 3). After the 800 s of this test, the voltage on the cell reached 5.2 V. The package of the battery was corrupted, and the battery was irrefutably damaged because of the consequent electrolyte leak. This state was also reached due to the inactivation of the safety valve located between battery terminals. Based on these experiments, the limitations of the given types of traction cells have been verified. Exceeding allowable charging voltage above 5 V causes non-reversible damage to the battery. Thus, it is not possible to regenerate it or restore it. The only way is to recycle it for the second use.

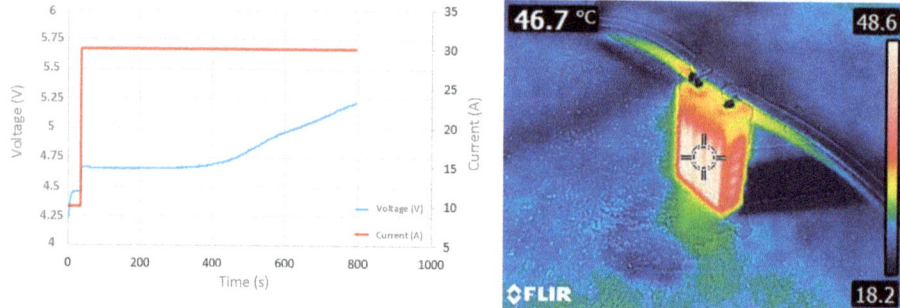

Figure 3. Waveforms of charging current and voltage during overcharging concerning 6.25 V of charging voltage level.

2.2. Experimental Application of Short-Circuit

The experiment of short-circuiting of selected $LiFePO_4$ 3.2 V, 40 Ah, 128 Wh battery cell was provided due to the requirement on the development of the recovery algorithm. Therefore, it was required to initially short-circuit selected cells for a given time interval in order to have a reference sample. From the safety point of view, short-circuit presents the most critical operational condition, because of the deformation of energy storage component that is caused by primarily—chemical and by secondarily—thermal issues. The experiment of the short-circuit was realized with the use of set-up shown in Figure 4. This configuration uses thermo-vision camera FLIR E5 and thermo-vision camera FLIR SC 660.

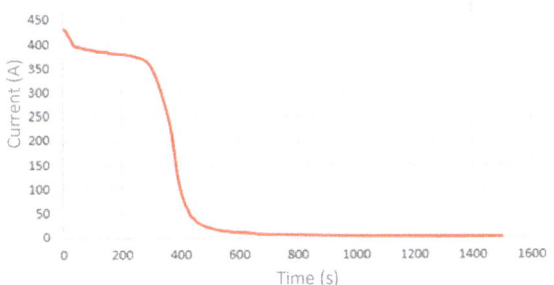

Figure 4. Time-waveform of short-circuit current during the experiment.

The utilization of both types is conditioned by the static and dynamic record of cell temperature. The measurement of short-circuit current was provided by APPA A18 (APPA Technology Corp., New Taipei City, Taiwan). Current meter, while values vs. time have been stored on PC through LabView measurement cards NI PXI 1031 (National Instruments, Austin, TX, USA). Camera Canon EOS 6D (Canon, Ota City, Tokio, Japan). was responsible for the acquirement of pictures in given time steps for the evaluation of cell's geometry shape changes.

Short-circuit was realized with the use of a new cell while it was initially formatted from the manufacturing process. Secondary formatting was done within laboratory conditions before the short-circuit experiment. The value of open-circuit voltage (OCV) before the test achieves 3.24 Vdc. For the start of the short-circuit test, the mechanical switch with a very high current rating was used.

Figure 4 shows the time-waveform of short-circuit current during the experiment. It is seen that after immediate shorting, the battery current reached over 430 A. Consequently, for more than 300 s, the battery was sourcing current over 350 A; its rapid drop is visible after 400 s of the short-circuit duration. The reduction to the value of 11.5 A and finally to 2.7 A was reached after more than 600 s. The total duration of this experiment was 25 min.

Together with the record of the value of short-circuit current, the thermal performance was also captured. At the end of the experiment, the surface temperature of the tested cell was 49.2 °C (Figure 5). The safety pressure valve located between the battery electrodes was not activated.

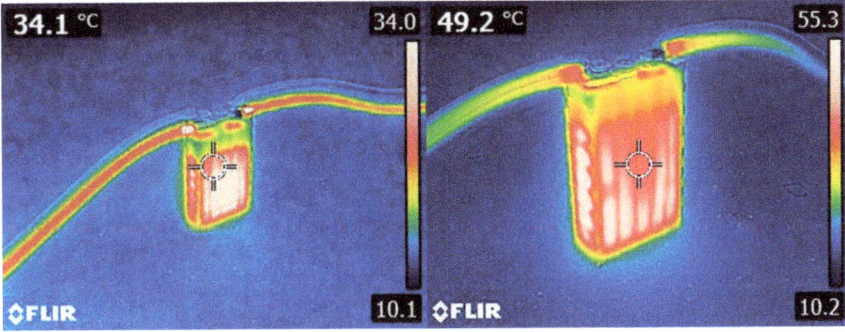

Figure 5. Maximal operational temperature of investigated battery during short-circuit operation.

Figure 6 indicates structural damages after completion of the short-circuit test. It is seen that the package of the cell is slightly flatulent. Geometrical measurements confirmed a visible increase of the width dimension from 46 mm up to 54 mm. The maximum of 54 mm was measured in the middle of the height of the battery package.

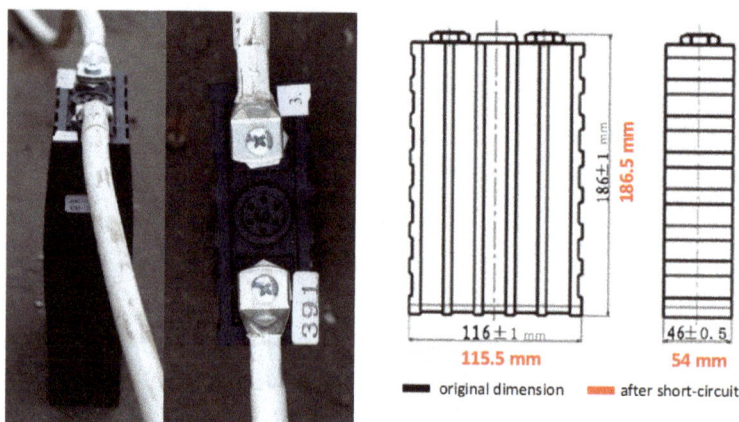

Figure 6. Geometrical changes of the cell package after short-circuit test.

After the experiment, the tested cell was left resting for 22 days. Dimensions of the package have been once checked after this period, while no change was observed compared to immediate evaluation after the short-circuit experiment.

2.3. Deep Discharge

Within the deep discharge testing, a 3.2 V 60 Ah LiFePO$_4$ cell was selected with significant damage to the cell's package. The measurement confirmed a deep discharge condition, while the open-circuit voltage of this cell was 1.88 V. The minimum voltage range for this cell is also 2.5 V, according to the datasheet (Table 1). An important fact is that the cell has a metal casing, which is primarily intended for increased protection against possible damage and heat dissipation. The tested cell represents an unused device, whereby deep discharge is a result of improper storage that has reached a critical level of voltage by self-discharge. It is determined by the manufacturer at a value of 3% of capacity over one month. Observation of the package of this cell discovers visible inflation within the central part, while the width reached 43.8 mm (Figure 7). The original cell's width stated by the manufacturer is 36 mm.

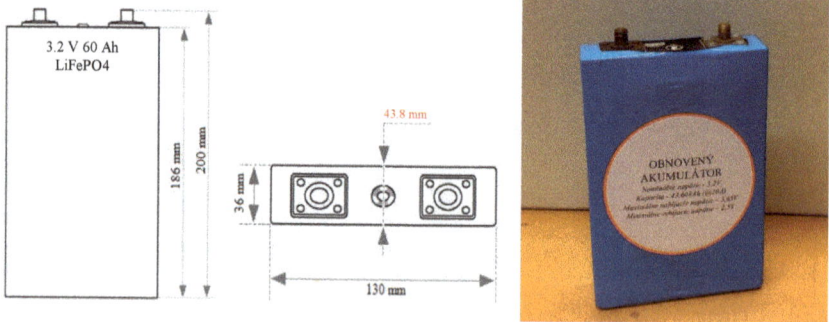

Figure 7. Evaluation of geometrical changes valid for center points of the width of deeply discharged cell (**left**) and a physical sample of this cell (**right**).

3. Regeneration Procedure Proposal for Short-Circuited and Deep Discharged Cells

The device under test is located within a metal box, while required measurements are realized with the use of laboratory equipment given on Figure 8. The mechanical switch protects the power line from the source/load to the battery preventing a hazardous situation. The programable electronic load KIKOSUI PLZ 100 W and programable DC source EA PSI 8080-60 are controlled by LabView interface, which also provides data logging of the measured values (cell's—current/voltage/temperature). Through a developed user guide, it is possible to program various scenarios related to recovery algorithms, i.e., charging sequences and discharging sequences.

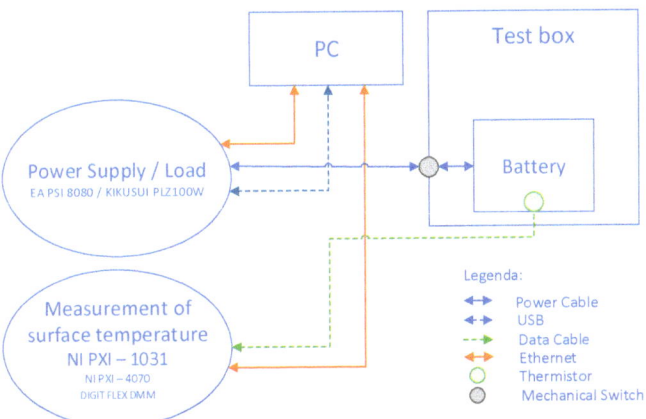

Figure 8. Block diagram of test-stand for traction batteries recovery.

3.1. Proposal for Regeneration Algorithm of Short-Circuited Cell

The proposed recovery process is suited for 3.2 V, 40 Ah, $LiFePO_4$, while its use is available for various Li-Fe phosphate cells. The only change lies in the consequent modification of individual charging steps, considering maximum allowable charging current and voltage. After the proposed recovery method is applied, its effect will be evaluated by the test of obtained Ah of the recovered cell.

The initial check of the harmed cell was focused on the evaluation of OCV, whose value was 2.83 V. This refers to the fully discharged state, while the value of OCV after long-term short circuit is still between the limits of operational values of the cell defined by the datasheet. For the selected type, the limits are within 2.5 V ± 3.65 V.

The battery recovery aims to reduce the negative impacts of the short-circuit consequences on the electrical and mechanical properties of DUT (device under test).

The proposal is based on the charging sequences, which are characterized by the pulsed current (Figure 9). These charging sequences are split into six groups. After each of the sequences, the resting period is applied (app. 16 h). The duration of each pulse sequence was lasting 100 min, whereby four cycles (25 min each) developed one sequence. The main difference between the cycles is the amplitude of charging current (Table 2).

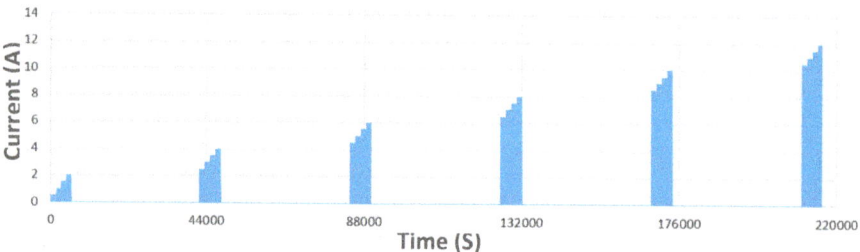

Figure 9. Graphical interpretation of proposed regeneration procedure for short-circuited cells including resting periods of DUT.

Table 2. The setting of individual sequences of regeneration algorithm for short-circuited cell.

Sequence	Duration	The Amplitude of Charging Current	Charging Voltage
1	100 min 4 cycles × 25 min	0.5 A–2 A each cycle 0.5 A increase	3.65 V
2	100 min 4 cycles × 25 min	2.5 A–4 A each cycle 0.5 A increase	3.65 V
3	100 min 4 cycles × 25 min	4.5 A–6 A each cycle 0.5 A increase	3.65 V
4	100 min 4 cycles × 25 min	6.5 A–8 A each cycle 0.5 A increase	3.65 V
5	100 min 4 cycles × 25 min	8.5 A–10 A each cycle 0.5 A increase	3.65 V
6	100 min 4 cycles × 25 min	10.5 A–12 A each cycle 0.5 A increase	3.65 V

Figure 10 shows the time waveform of the battery voltage and temperature during the first charging sequence. It is seen that the initial OCV value was 2.83 V, while at the end of the charging sequence, the value reached 2.97 V.

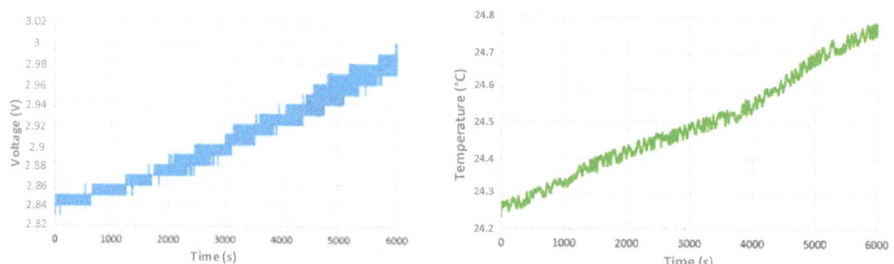

Figure 10. Time waveform of battery voltage (**left**) and its surface temperature (**right**) during the first regeneration sequence.

After the last 16 h of regeneration, the last sequence was applied, while OCV drops to 3.26 V what is a difference of 0.11 V compared to the end of the fifth charging sequence (Figure 11). The current range within the sixth cycle is 10.5 A up to 12 A. The value of the battery voltage at the end was 3.42 V, while after final 16 h of resting period OCV reached 3.31 V. The temperature profile during each regeneration sequence has shape similar to characteristics shown on Figures 10 and 11 as well.

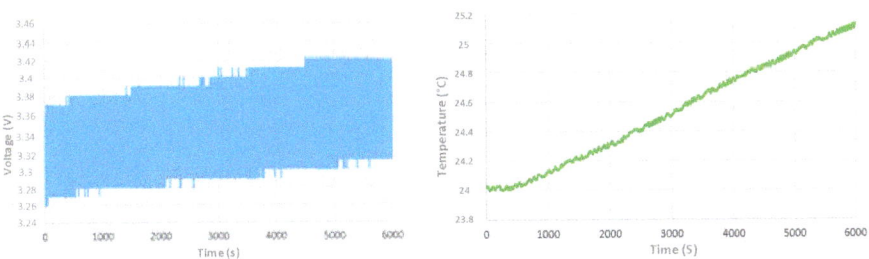

Figure 11. Time waveform of battery voltage (**left**) and its surface temperature (**right**) during last regeneration sequence.

DUT has completed six charging sequences. After 18 h of regeneration from the last sixth charging sequence, the open-circuit voltage reached 3.27 V. From the viewpoint of the safety of cell operation during the application of the charging sequences, no significant increase in the surface temperature or change in the dimensions of the external structure was observed.

The main reason for the cycling of each subsequence was to ensure a gradual increase in the charging current. The pause interval between charging steps is essential for battery recovery. In terms of reliability, the battery is currently stable and ready for a full charge. The level of its usability will be evaluated based on the test of delivered ampere-hours.

3.2. Proposal for Regeneration Algorithm of Deeply Discharged Cell

As discussed earlier, the 3.2 V 60 Ah, LiFePO$_4$ cell was selected for the application of a regeneration algorithm while deep discharge (1.88 V) of the DUT is considered. Compared to the regeneration of the short-circuited cell, this algorithm is divided into 30 charging cycles and 30 cycles of pause mutually alternating, whereby one cycle is lasting 1 s (Table 3). After it, 5 min of regeneration is applied to DUT (Figure 12). One sequence of regeneration algorithm valid for deep discharge was lasting 126 min. The charging current impulse had an amplitude of 20 A, which refers to 1/3 of the capacity of the cell. The amplitude of the charging voltage was 3.65 V. For the given battery, the application of six sequences was realized in order to achieve the required OCV on the device. At the same time, 16 h of the resting period was applied between individual sequences.

Table 3. The setting of sequences of regeneration algorithm for deeply discharged cell.

	Duration	The Amplitude of Charging Current	Charging Voltage
Sequence	126 min consists of subsequences 1 and 2	20 A	3.65 V
Subsequence 1	30 charging pulses 30 pause pulses alternating one pulse = 1 s		
Subsequence 2	Regeneration period 5 min		

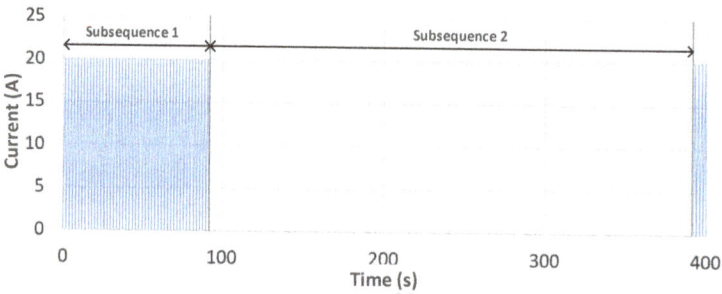

Figure 12. Graphical interpretation of the proposed regeneration sequence for deeply discharged cells.

Figure 13 shows the time waveform of battery voltage during the application of the first regeneration sequence and last (six) regeneration sequence (Table 4). It is seen that voltage raised from 2.04 V up to 3.19 V at the end of the first sequence. During the last sequence, the voltage level on the cell exceeds 3.3 V. The temperature on the cell during each sequence was within 25.38 °C–26.18 °C.

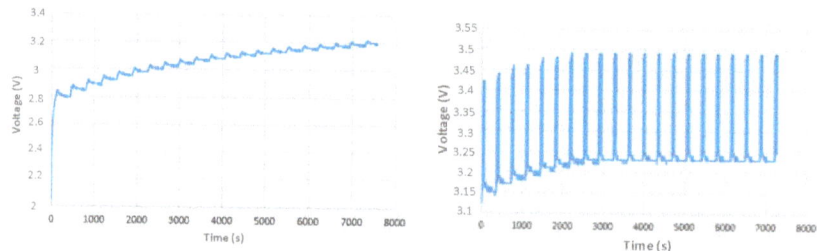

Figure 13. The voltage waveform of initially deeply discharged cell after application of the first sequence of regeneration algorithm (**left**) and the last, sixth sequence (**right**).

Table 4. Voltage levels before and after each regeneration sequence of deeply discharged cell.

Sequence	The Voltage on the Cell before the Sequence	The Voltage on the Cell after the Sequence
1	2.04 V	3.19 V
2	3.12 V	3.21 V
3	3.19 V	3.27 V
4	3.21 V	3.28 V
5	3.24 V	3.29 V
6	3.26 V	3.31 V

4. Verification of Recovered Cells Through the Test of Delivered Ampere-Hours

At the beginning of this test, it is required to charge recovered batteries fully. For selected types of batteries, CC&CV charging (Constant Current and Constant Voltage) is recommended. Both recovered batteries are verified in the way of delivered ampere-hours test (test of capacity), whereby new un-damaged cells have been used as reference devices for comparisons and evaluation.

For the test of battery capacity, five discharging scenarios have been verified. Each scenario is characterized by a different value of discharging current, while the range was selected based on the operational properties of selected cells (13 A/20 A–120 A). After each test, the cell was re-charged to full capacity.

4.1. Verification of Regeneration Algorithm of Short-Circuited Cell

Test of the capacity of the short-circuited cell was realized for five values of discharging currents, i.e., 13 A, 20 A, 40 A, 80 A, and 120 A. Initially recovered cell was tested. At the same time, consequently, the reference sample has undergone a similar test. The profiles of battery voltage during individual discharging states for the recovered and new cell are graphically interpreted on Figures 14–16.

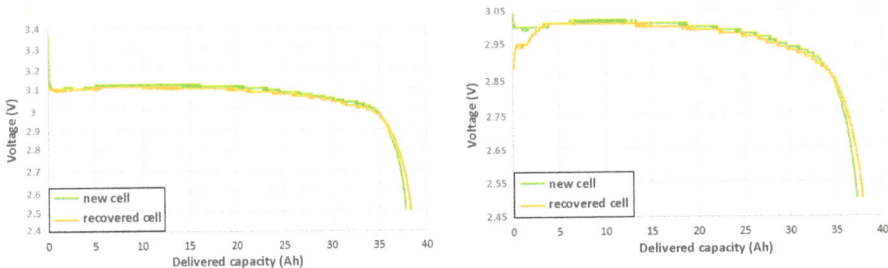

Figure 14. Voltage profile during discharge by 13 A (**left**) and 20 A (**right**) for regenerated (yellow) and referenced cell (green).

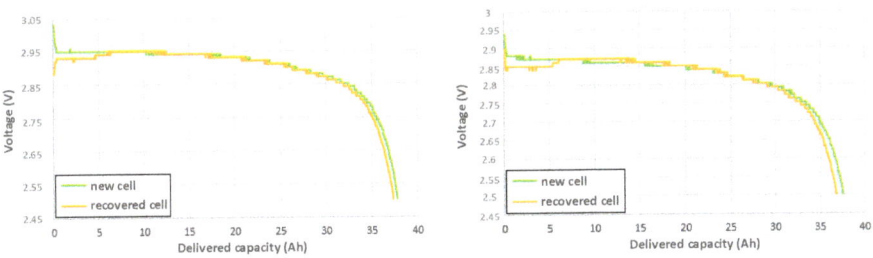

Figure 15. Voltage profile during discharge by 40 A (**left**) and 80 A (**right**) for regenerated (yellow) and referenced cell (green).

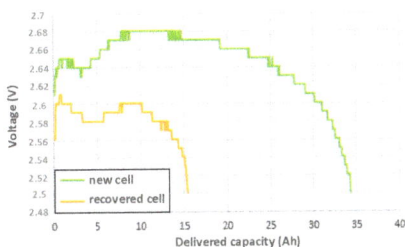

Figure 16. Voltage profile during discharge by 120 A for regenerated (yellow) and referenced cell (green).

From Figures 14 and 15 is visible that for discharging current between 13 A–80 A recovered cell is delivering a similar number of ampere-hours compared to the new cell. The visible difference is valid for the case of 120 A (Figure 16), where the recovered cell provides just half of the capacity of the new cell. A detailed summary of the results from this test is listed in Table 5.

Table 5. Summary of the results of the verification test of short-circuited cell.

Battery Cell Model	3.2 V, 40 Ah, LiFePO$_4$	3.2 V, 40 Ah, LiFePO$_4$
Cell Status	Recovered after Short-Circuiting	New Cell
Discharge CC 20 A		
Discharge time	1 h, 52 min, 54 s	1 h, 51 min
Ambient temperature	20.105 °C	21.451 °C
Maximal surface temperature	28.1 °C	30.253 °C
Delivered Ah	38.297 Ah	37.752 Ah
Discharge CC 40 A		
Discharge time	55 min, 44 s	54 min, 26 s
Ambient temperature	20.677 °C	21.054 °C
Maximal surface temperature	29.684 °C	31.374 °C
Delivered Ah	37.812 Ah	37.163 Ah
Discharge CC 60 A		
Discharge time	36 min, 48 s	36 min, 52 s
Ambient temperature	21.264 °C	21.984 °C
Maximal surface temperature	32.86 °C	33.036 °C
Delivered Ah	37.303 Ah	37.726 Ah
Discharge CC 80 A		
Discharge time	27 min, 41 s	27 min, 34 s
Ambient temperature	20.384 °C	21.453 °C
Maximal surface temperature	34.788 °C	34.346 °C
Delivered Ah	36.755 Ah	37.482 Ah
Discharge CC 120A		
Discharge time	7 min, 47 s	16 min, 44 s
Ambient temperature	21.116 °C	20.998 °C
Maximal surface temperature	28.186 °C	35.027 °C
Delivered Ah	15.313 Ah	34.236 Ah

4.2. Verification of Regeneration Algorithm of Deeply Discharged Cell

The second verification test of the recovery algorithm for the deeply discharged cell was realized for four values of discharging currents, i.e., 20 A, 40 A, 60 A, and 80 A. Recovered cell was compared with unused (new) cell, which was initially formatted.

Figure 17 shows the voltage profile of recovered and new cells for 20 A and 40 A of discharging current. For 20 A situation, the new cell delivers an app. 60 Ah, more precisely 60.869 Ah during 3 h 2 min 46 s. The highest temperature on the surface of the cell achieved 35.397 °C. On the other side, the recovered cell delivers just 43.608 Ah what is more than 17 Ah less compared to the new cell. For the test with 40 A of discharging current, the new cell delivered 59.468 Ah and recovered 42,536 Ah. During both tests, the temperature on the surface of the cell raised to 39.16 °C, i.e., 3.46 °C more compared to the recovered cell.

Voltage profiles for the tests with 60 A and 80 A are shown on Figure 18. It is seen that for both situations, the new cell can deliver approximately 56 Ah (surface temperature 44.168 °C). In contrast, recovered cell behaves similar to previous tests, if 60 A discharge is considered (app. 40 Ah is delivered). However, if 80 A of discharge is applied to the recovered cell, its voltage drops to the minimum cell voltage. Thus, this situation almost represents the hazardous case. The amount of delivered Ah reached 35.886 Ah (surface temperature 40.881 °C). The summary of all tests is listed in Table 6.

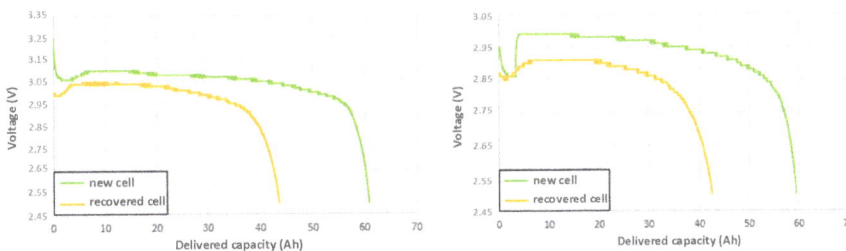

Figure 17. Voltage profile during discharge by 20 A (**left**) and 40 A (**right**) for regenerated (yellow) and referenced cell (green).

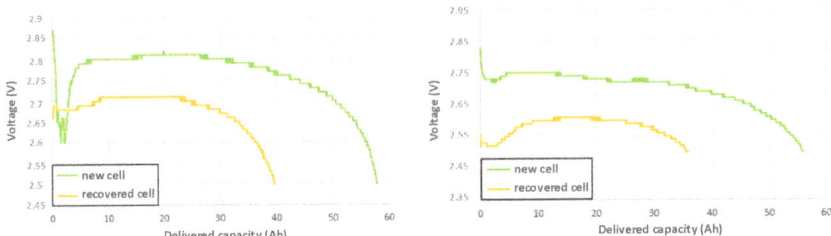

Figure 18. Voltage profile during discharge by 60 A (**left**) and 80 A (**right**) for regenerated (yellow) and referenced cell (green).

Table 6. Summary of the results of the verification test of deeply discharged cell.

Battery Cell Model	3.2 V, 60 Ah, LiFePO$_4$	3.2 V, 60 Ah, LiFePO$_4$
Cell Status	Recovered after Deep Discharge	New Cell
Discharge CC 20 A		
Discharge time	2 h, 8 min, 18 s	3 h, 2 min, 46 s
Ambient temperature	21.238 °C	22.018 °C
Maximal surface temperature	31.293 °C	35.397 °C
Delivered Ah	43.608 Ah	60.869 Ah
Discharge CC 40 A		
Discharge time	1 h, 2 min, 40 s	1 h, 27 min, 18 s
Ambient temperature	21.896 °C	21.997 °C
Maximal surface temperature	35.703 °C	39.158 °C
Delivered Ah	42.536 Ah	59.486 Ah
Discharge CC 60 A		
Discharge time	38 min, 58 s	57 min, 47 s
Ambient temperature	20.891 °C	22.321 °C
Maximal surface temperature	38.403 °C	41.502 °C
Delivered Ah	39.709 Ah	57.826 Ah
Discharge CC 80 A		
Discharge time	26 min, 22 s	41 min, 4 s
Ambient temperature	21.574 °C	22.054 °C
Maximal surface temperature	40.881 °C	44.168 °C
Delivered Ah	35.886 Ah	55.965 Ah

5. Conclusions

In this paper, the experimental investigation of the recovery algorithms of the traction batteries with lithium phosphate technology has been studied with selected types of cells. The main focus was given on possibilities related to the renewal of initially damaged cells by long-term short-circuit or deep discharge. For both situations, procedures are based on the charging process. Thus, individual procedures propose a different approach, i.e., the short-circuited cell requires gradual medium duration profile of charging sequences (with an increase of charging current), whereby deeply discharged cell uses short duration peak charging pulses. Both proposals have been experimentally verified in the way of the test of delivered ampere-hours of restored cells. The comparisons of these tests have been made with newly formatted cells of the same type. From experiments was found that short-circuited battery is capable of recovering up to 80% if the proposed recovery procedure is applied. It is valid for discharge currents within 20 A–80 A (0.5 °C–2 °C). For higher currents, the recovery represents an app. 55%. From these results can be said, that recovered cell after long-term short-circuit is capable of second use. At the same time, restrictions must be respected related to the value of continuous discharging current. A similar result was achieved if the deeply discharged cell was verified. The recovery achieved almost 70% if discharge currents are within 0.3 °C–0.6 °C. For higher currents, the voltage drop of the battery represents limiting parameters as it reaches the minimum allowable operational value.

Batteries were tested with an ampere-hour test after the initial testing sequence ten times in a row after 28 days of storage. Differences in capacity values between repeated tests were lower than 2%. From received results, the expectations for long-term regenerated battery use is possible regarding recovered capacity, which is lower than nominal capacity. Restrictions on discharging currents must also be accepted, i.e., it is not recommended to use high operating currents. Consequently, it is proposed to use accelerated pulsed charging instead of the CV/CC method to slow down degradation lengthening operational life.

It must be said here that the same results have been achieved after 28 days of these tests. The proposed methodology gives proper way, how to recover damaged lithium cells.

Author Contributions: Conceptualization, P.S., methodology, J.A., experimental analysis and set-ups, M.D.; writing—original draft preparation, M.F. All authors have read and agreed to the published version of the manuscript.

Funding: This research was funded by APVV-15-0396 and APVV-15-0571. The experimental support was also done by project funding Vega 1/0547/18.

Acknowledgments: The authors would like to thank to Slovak national grant agencies APVV and Vega for the above mentioned financial support.

Conflicts of Interest: The authors declare no conflicts of interest.

References

1. Propfe, B.; Redelbach, M.; Santini, D.J.; Friedric, H. Cost Analysis of Plug-In Hybrid Electric Vehicles Including Maintenance & Repair Costs and Resale Values (Conference EVS26). 2012. Available online: https://www.mdpi.com/2032-6653/5/4/886 (accessed on 25 May 2020).
2. Nykvist, B.; Nilsson, M. Rapidly falling costs of battery packs for electric vehicles. *Nat. Clim. Chang.* **2015**, *5*, 329. [CrossRef]
3. Scrosati, B. History of lithium batteries. *J. Solid State Electrochem.* **2011**, *15*, 1623–1630. [CrossRef]
4. Brodd, R. *Batteries for Sustainability: Selected Entries from the Encyclopedia of Sustainability Science and Technology*; Springer: New York, NY, USA, 2013; Volume VI, p. 513. ISBN 978-1-4614-5791-6.
5. Aditya, P.J.; Ferowsi, M. Comparison of NiMh and Li-ion batteries in automotive applications. In Proceedings of the Vehicle Power and Propulson Conference, Harbin, China, 3–5 September 2008.
6. Yoo, H.D.; Markevich, E.; Salta, G.; Sharon, D.; Aurbach, D. On the challenge of developing advanced technologies for electrochemical energy storage and conversion. *Mater. Today* **2014**, *17*, 110–121. [CrossRef]
7. Tarascon, J.M.; Armand, M. Issues and challenges facing rechargeable lithium batteries. *Nature* **2001**, *414*, 359–367. [CrossRef] [PubMed]
8. Becker, J.; Schaeper, C. Design of a safe and reliable li-ion battery system for applications in airbone system. In Proceedings of the 52nd AIAA Aerospace Sciences Meeting-AIAA Science and Technology Forum and Exposition, National Harbor, MD, USA, 13–17 January 2014; SciTech: Wellesley Park, NC, USA, 2014.

9. Chamnan-arsa, S.; Uthaichana, K.; Kaewkham-ai, B. Modeling of LiFePO$_4$ battery state of charge with recovery effect as a three-mode switched system. In Proceedings of the 2014 13th International Conference on Control, Automation Robotics & Vision (ICARCV), Singapore, 10–12 December 2014; pp. 1712–1717. [CrossRef]
10. Cai, Y.; Zhang, Z.; Zhang, Y.; Liu, Y. A self-reconfiguration control regarding recovery effect to improve the discharge efficiency in the distributed battery energy storage system. In Proceedings of the 2015 IEEE Applied Power Electronics Conference and Exposition (APEC), Charlotte, NC, USA, 15–19 March 2015; pp. 1774–1778. [CrossRef]
11. Huang-Jen, C.; Li-Wei, L.; Ping-Lung, P.; Ming-Hsiang, T. A novel rapid charger for lead-acid batteries with energy recovery. *IEEE Trans. Power Electron.* **2006**, *21*, 640–647. [CrossRef]
12. Stefan-Cristian, M.; Cornelia, C.A.; Stefan, U. Battery regeneration technology using dielectric method. In Proceedings of the 2014 International Conference and Exposition on Electrical and Power Engineering (EPE), Iasi, Romania, 16–18 October 2014; pp. 839–844. [CrossRef]
13. Orchard, M.E.; Lacalle, M.S.; Olivares, B.E.; Silva, J.F.; Palma-Behnke, R.; Estévez, P.A.; Cortés-Carmona, M. Information-Theoretic Measures and Sequential Monte Carlo Methods for Detection of Regeneration Phenomena in the Degradation of Lithium-Ion Battery Cells. *IEEE Trans. Reliab.* **2015**, *64*, 701–709. [CrossRef]
14. Amanor-Boadu, J.M.; Guiseppi-Elie, A. Improved Performance of Li-ion Polymer Batteries Through Improved Pulse Charging Algorithm. *Appl. Sci.* **2020**, *10*, 895. [CrossRef]
15. Kirpichnikova, I.; Korobatov, D.; Martyanov, A.; Sirotkin, E. Diagnosis and restoration of li-Ion batteries. *J. Phys. Conf. Ser.* **2017**. [CrossRef]
16. Medora, K.N.; Kusko, A. An Enhanced Dynamic Battery Model of Lead-Acid Batteries Using Manufacturers Data. In Proceedings of the INTELEC 06—Twenty-Eighth International Telecommunications Energy Conference, Providence, RI, USA, 10–14 September 2006; pp. 1–8. [CrossRef]
17. Cacciato, M.; Nobile, G.; Scarcella, G.; Scelba, G. Real-Time Model-Based Estimation of SOC and SOH for Energy Storage Systems. *IEEE Trans. Power Electron.* **2017**, *32*, 794–803. [CrossRef]
18. Liu, J.; Chen, Z. Remaining Useful Life Prediction of Lithium-Ion Batteries Based on Health Indicator and Gaussian Process Regression Model. *IEEE Access* **2019**, *7*, 39474–39484. [CrossRef]
19. Baghdadi, I.; Briat, O.; Deletage, J.Y.; Vinassa, J.M.; Gyan, P. Dynamic Battery Aging Model: Representation of Reversible Capacity Losses Using First Order Model Approach. In Proceedings of the 2015 IEEE Vehicle Power and Propulsion Conference (VPPC), Montreal, QC, Canada, 19–22 October 2015; pp. 1–4. [CrossRef]

© 2020 by the authors. Licensee MDPI, Basel, Switzerland. This article is an open access article distributed under the terms and conditions of the Creative Commons Attribution (CC BY) license (http://creativecommons.org/licenses/by/4.0/).

Article

A Fractional-Order Kinetic Battery Model of Lithium-Ion Batteries Considering a Nonlinear Capacity

Qi Zhang [1,2], Yan Li [1], Yunlong Shang [1], Bin Duan [1,2], Naxin Cui [1,*] and Chenghui Zhang [1,*]

1. School of Control Science and Engineering, Shandong University, Jinan 250061, China; zhangqi2013@sdu.edu.cn (Q.Z.); liyan_cse@sdu.edu.cn (Y.L.); shangyunlong@mail.sdu.edu.cn (Y.S.); duanbin@sdu.edu.cn (B.D.)
2. State Key Laboratory of Automotive Simulation and Control, Jilin University, Changchun 130025, China
* Correspondence: zchui@sdu.edu.cn (C.Z.); cuinx@sdu.edu.cn (N.C.); Tel.: +86-0531-88392907

Received: 9 March 2019; Accepted: 28 March 2019; Published: 2 April 2019

Abstract: Accurate battery models are integral to the battery management system and safe operation of electric vehicles. Few investigations have been conducted on the influence of current rate (C-rate) on the available capacity of the battery, for example, the kinetic battery model (KiBaM). However, the nonlinear characteristics of lithium-ion batteries (LIBs) are closer to a fractional-order dynamic system because of their electrochemical materials and properties. The application of fractional-order models to represent physical systems is timely and interesting. In this paper, a novel fractional-order KiBaM (FO-KiBaM) is proposed. The available capacity of a ternary LIB module is tested at different C-rates, and its parameter identifications are achieved by the experimental data. The results showed that the estimated errors of available capacity in the proposed FO-KiBaM were low over a wide applied current range, specifically, the mean absolute error was only 1.91%.

Keywords: kinetic battery model; lithium-ion batteries; nonlinear capacity; fractional calculus

1. Introduction

Electric vehicles have the advantages of high fuel economy and zero exhaust emissions [1,2]. As clean and efficient energy sources, power batteries are core components and are critical to the comprehensive performance of vehicles. Lithium-ion batteries (LIBs) show strong overall advantages in the field of power batteries because of their high energy density, long life, and excellent cycle performance [2,3].

Battery states mainly include the state of charge (SOC), state of health (SOH), state of power (SOP), state of energy (SOE), and state of function (SOF) [4–6]. However, they cannot be measured directly; they can only be estimated by testing battery voltage, current, and temperature, among other factors. Battery state estimation is extremely important in battery management systems (BMSs) to guarantee the safe and reliable operation of batteries, and a multitude of research has investigated the methods of state estimation of LIBs based on an accurate model [7–14]. Commonly used battery models include electrochemical models (EchMs) [11,14], analytical models (AMs) [7,12–14], stochastic models (SMs) [7,12,14], neural network models (NNMs) [14], and equivalent circuit models (ECMs) [8–11,14]. EchMs are accurate in describing the internal electrochemical reaction using complex, nonlinear differential equations, but they are difficult to understand. AMs model the major properties of the battery using only a few equations, and are much easier to use than EchMs. SMs mainly concern the battery recovery characteristics as a Markov process, in which the pulse discharge characteristics of the battery can be described, but they are not applicable for variable current. NNMs have fast parallel processing capabilities as well as strong self-learning and self-organizing abilities, but they require a large amount of training data, and errors can arise from the training data and training methods. ECMs are widely used for electrical design and modelling simulations [9,10] because they can accurately describe the battery voltage-current (U-I) performance. Many improved models have been proposed by scholars that not only describe U-I performance (external characteristics) but also the capacity performance (internal characteristics). For instance, an ECM with a variable effective capacity for LIBs is proposed in [15], and the model is further optimized using computational intelligence techniques [16].

However, an accurate and concise battery model is not readily achievable because the models are highly nonlinear and complex. Few investigations have been conducted on the influence of the current rate (C-rate) on the available capacity of batteries because the characteristics of LIBs will change significantly under different conditions in view of its sensitivity to C-rate, temperature, cycle life, etc. As described in detail in Figure 1, the battery capacity is not similar to water in a bucket. It is mainly manifested as "capacity nonlinear effect" and "recovery effect" [7,11,12]. The capacity nonlinear effect is that the available capacity will decline nonlinearly with C-rates; the greater the discharge current, the less the available capacity. The recovery effect is that the battery's available capacity will rise up when discharge is stopped. For instance, a released capacity of 50 A will be much less than that at 5 A, and the capacity will be restored if the battery rests for a while.

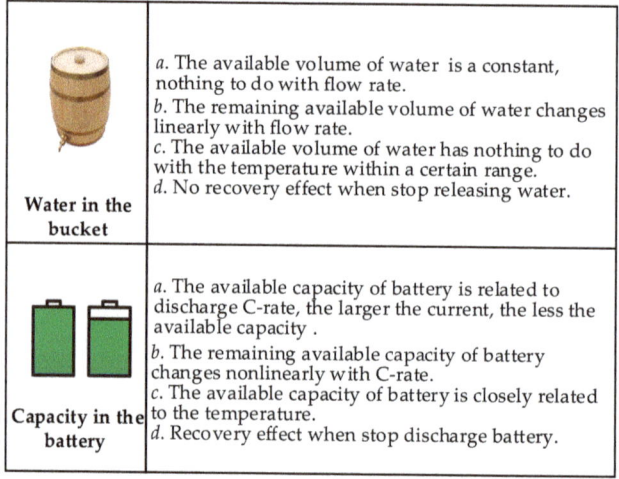

Figure 1. Battery capacity is not similar to water in a bucket.

The remaining available battery capacity is critical in electric vehicles, similar to the role of the remaining fuel in internal combustion vehicles. Thus, the available capacity estimation of a battery considering C-rate is very important. Overall, commonly used models for available capacity estimation of the battery mainly include two classic analytical models: Peukert's law and the kinetic battery model (KiBaM). Proposed by Wilhelm Peukert, Peukert's law was used to estimate nonlinear delivered capacity and predict the battery run time of a rechargeable lead–acid battery at different constant discharge C-rates from the fully charged state [17]. The nonlinear properties between the available capacity and the C-rate are considered, and the battery's run time can be approximated. The model itself is relatively simple compared to KiBaM. However, it does not consider the recovery effect of the battery when discharge stops. Fortunately, Manwell and McGowan proposed a kinetic battery model (KiBaM) to model lead–acid storage batteries in 1993 [13]. It is intuitionistic and easy to understand based on perceptual knowledge. Moreover, it can be used in modelling and simulation [7,12–14]. In [14], a widely-used KiBaM was used to capture nonlinear capacity effects for accurate SOC tracking and runtime predictions of the battery. In [18], KiBaM was extended to consider the temperature effect on battery capacity. The proposed temperature-dependent KiBaM (T-KiBaM) can handle operating temperatures, and it can provide better estimates for battery lifetimes and voltage behaviors. However, the classic KiBaM is described by regular calculus. A precise and concise battery model at various conditions has always been challenging for researchers to create. Using fractional calculus with impedance models is quite common in the modelling of energy storage and generation elements, including capacitors/super capacitors [19–21] and batteries [19,22,23]; what is more, the fractional calculus has also been used in the state estimation and prediction of batteries [21,23,24]. Actually, the nonlinear characteristics of lithium-ion batteries are closer to a fractional-order dynamic system, because the material diffusion and electrochemical properties have been successfully described using fractional calculus. The application of fractional-order models to represent physical systems is timely and interesting.

In this paper, a novel fractional-order KiBaM (FO-KiBaM) is proposed to describe the nonlinear capacity characteristics of LIBs. The research ideas and arrangement of the rest of the paper are as follows. Firstly, the capacity nonlinear effect and recovery effect of KiBaM are analyzed in Section 2. The basic principle of fractional calculus and its application in the proposed FO-KiBaM are introduced in Section 3. In Section 4 the charge and discharge experiments of a LIB module are designed and conducted under different C-rates. Finally, the results and model error analyses are compared and illustrated in Section 5, followed by the conclusion in Section 6.

2. Kinetic Battery Model (KiBaM)

As shown in Figure 2, KiBaM uses two wells of different sizes to describe the dynamic changes in battery capacity. Thus, it is also called a "two-well" model. The two wells represent the "directly available capacity" and the "temporary capacity" of the battery, respectively [13,14]. The directly available capacity can be obtained directly at discharge, denoted as y_1, and its height is denoted as h_1. The "temporary capacity" cannot be directly obtained at discharge, denoted as y_2, and its height is denoted as h_2. It is easy to understand that the sum of y_1 and y_2 is the total capacity of the battery, and the sum of y_1 and a part of y_2 is the available capacity of the battery. The letter k represents the rate that the charge flow from y_2 into y_1; and the letter c represents the capacity proportion of the two wells. The k and c variables affect the nonlinear battery capacity characteristics; notably, they are closely related to the materials and composition of the battery.

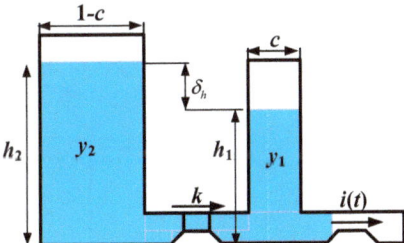

Figure 2. Kinetic battery model (KiBaM).

It is clear that the directly available capacity y_1, the temporary capacity y_2, and the heights h_1 and h_2 in the KiBaM are satisfied by the following:

$$\begin{cases} h_1(t) = \frac{y_1(t)}{c}, \\ h_2(t) = \frac{y_2(t)}{1-c}, \\ \delta_h(t) = \frac{y_2(t)}{1-c} - \frac{y_1(t)}{c}, \end{cases} \quad (1)$$

where δ_h represents the height difference between the two wells.

The principle of the KiBaM is as follows. When the battery is discharged, the charge of y_1 flows out, simultaneously, the charge of y_2 flows into y_1 slowly with k. The charge flowing out of y_1 is faster than flowing in, so the height difference between y_1 and y_2 will increase. The larger the discharge current, the less capacity that is released, which reflects the nonlinear effect of the battery's capacity. Additionally, the unavailable capacity of the battery is the capacity represented by the height differences of y_1 and y_2. Once discharging is stopped, the charge of y_2 will flow into y_1 slowly until the heights of y_1 and y_2 are equal, and the charge of y_1 will pick up, which reflects the battery's recovery effect. From the above intuitive graphical description of KiBaM, the changes of the charges y_1 and y_2 in the two wells can be expressed as follows:

$$\begin{cases} \frac{dy_1(t)}{dt} = -i(t) + k\delta_h(t) = -i(t) + k[\frac{y_2(t)}{1-c} - \frac{y_1(t)}{c}], \\ \frac{dy_2(t)}{dt} = -k\delta_h(t) = -k[\frac{y_2(t)}{1-c} - \frac{y_1(t)}{c}]. \end{cases} \quad (2)$$

From here, the capacity variance of the battery with time can be expressed as follows:

$$\begin{cases} C_{rem}(t) = y_1(t) + y_2(t), \\ C_{unav}(t) = (1-c)\delta_h(t), \\ C_{av}(t) = C_{rem}(t) - C_{unav}(t), \end{cases} \quad (3)$$

where C_{rem}, C_{unav}, and C_{av} represent the remaining capacity of battery, the unavailable capacity of battery, and the available capacity of battery, respectively.

Take the constant-current discharge for example. As shown in Figure 3, the time course can be described as follows:

$$i(t) = \begin{cases} I, t_0 \leq t \leq t_d, \\ 0, t_d < t \leq t_r, \end{cases} \quad (4)$$

where t_0, t_d, t_r represents the initial time, the discharge end time, and the recovery end time, respectively.

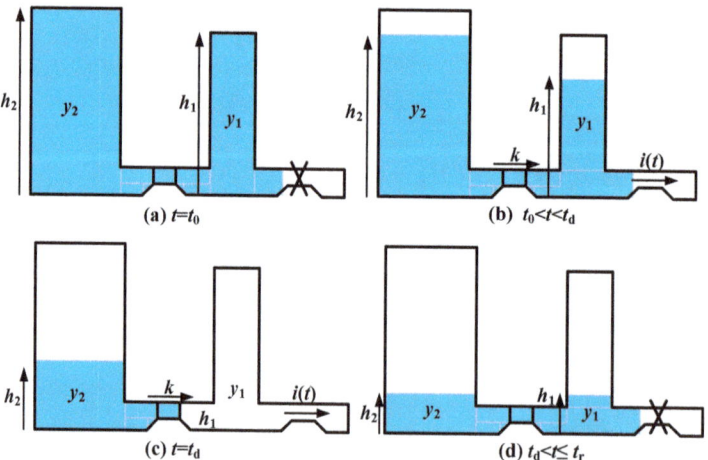

Figure 3. Battery capacity changes with the discharge time in the discharge process.

Actually, when the battery is discharged in a time-varying current, the entire process can be divided into multiple discharge segments in accordance with the time interval. In each segment, the battery can be considered discharged with a constant current and then set aside for a while.

By the Laplace transform and inverse Laplace transform of (2), the following expression can be obtained:

$$\begin{cases} y_1(t) = y_1(t_0)e^{-k'(t-t_0)} + \frac{(y_0k'c-I)[1-e^{-k'(t-t_0)}]}{k'} - \frac{Ic[k'(t-t_0)-1+e^{-k'(t-t_0)}]}{k'}, \\ y_2(t) = y_2(t_0)e^{-k'(t-t_0)} + y_0(1-c)[1-e^{-k'(t-t_0)}] - \frac{I(1-c)[k'(t-t_0)-1+e^{-k'(t-t_0)}]}{k'}, \end{cases} \quad (5)$$

where $k' = \frac{k}{c(1-c)}$.

The unavailable capacity of battery in (3) can be derived as follows:

$$C_{unav}(t) = \begin{cases} (1-c)[\delta_h(t_0)e^{-k'(t-t_0)} + \frac{I}{c}\frac{1-e^{-k'(t-t_0)}}{k'}], t_0 \leq t \leq t_d, \\ (1-c)\delta_h(t_d)e^{-k'(t-t_d)}, t_d < t \leq t_r. \end{cases} \quad (6)$$

It can be further expressed as follows:

$$C_{unav}(t) = \begin{cases} C_{unav}(t_0)e^{-k'(t-t_0)} + (1-c)\frac{I}{c}\frac{1-e^{-k'(t-t_0)}}{k'}, t_0 \leq t \leq t_d, \\ C_{unav}(t_d)e^{-k'(t-t_d)}, t_d < t \leq t_r. \end{cases} \quad (7)$$

As can be seen from (7), in the discharge period $t_0 \leq t \leq t_d$, the battery's unavailable capacity is impacted nonlinearly by the discharge time; the longer the time and the larger the current, the larger the unavailable capacity. In the stationary period $t_d < t \leq t_r$, the unavailable capacity decreases with time, because the charge of y_2 flows into y_1, which reflects the recovery effect of the battery in an open circuit state.

It can be seen that when $y_1 = 0$ (or $h_1 = 0$), the battery is discharged completely. At this point, all the remaining capacity is unavailable. Therefore, judging whether the battery is fully discharged is shown as follows:

$$C_{rem}(t) = C_{unav}(t) = (1-c)\delta_h(t). \quad (8)$$

The remaining available capacity of the battery can be expressed as follows:

$$C_{av}(t) = C_{t0} - \int_{t0}^{t} i(t)dt - C_{unav}(t) = C_{t0} - \int_{t0}^{t} i(t)dt - (1-c)\delta_h(t). \quad (9)$$

If the initial conditions $t_0 = 0$, and the battery is discharged at a constant current I, the initial state of y_1 and y_2 of the battery are shown as below:

$$\begin{cases} y_1(t_0) = y_1(0) = cC_0, \\ y_2(t_0) = y_2(0) = (1-c)C_0, \end{cases} \quad (10)$$

where C_0 is the initial total capacity of battery. Then, Formula (3) can be simplified as follows:

$$\begin{cases} C_{rem}(t) = C_0 - It, \\ C_{unav}(t) = (1-c)\delta_h(t) = (1-c)\frac{I}{c}\frac{1-e^{-k't}}{k'}, \\ C_{av}(t) = C_0 - It - (1-c)\frac{I}{c}\frac{1-e^{-k't}}{k'}. \end{cases} \quad (11)$$

3. The Proposed Fractional-Order KiBaM (FO-KiBaM)

3.1. Fractional Calculus Theory

Fractional calculus is not a new concept, in fact, it can be traced back to the discussion of Leibniz and Hospital's research work in 1695, but due to a variety of calculation difficulties in practical applications, it was only a purely theoretical exploration in the early days. However, fractional calculus attracts great attention in complex engineering applications because of the continuous developments in natural science and computer technology. In recent decades, theoretical and mathematical tools have been used in research in a multitude of disciplines, and they are especially successful in high-energy physics, fluid mechanics, viscoelastic material mechanics, anomalous diffusion, electronic components analysis, and system control [20,25]. Fractional calculus has been a research hotspot for its unique and irreplaceable advantages.

Fractional derivative definitions (FDD) are defined in different ways; the most commonly used include the Grunwald–Letnikov definition (GL-FDD), the Riemann–Liouville definition (RL-FDD), and the Caputo definition (Caputo-FDD) [20,25]. The GL-FDD is expressed as:

$$^{G}D_{t}^{\alpha}f(t) = \lim_{h \to 0} h^{-\alpha} \sum_{j=0}^{[\frac{t-t_0}{h}]} (-1)^{-j} \binom{\alpha}{j} f(t - jh), \tag{12}$$

where $^{G}D_{t}^{\alpha}$ represents the GL-FDD type; $f(t)$ is an arbitrary integrable function; α is an arbitrary real number; $n = [\frac{t-t_0}{h}]$ represents the integer part; and $\binom{\alpha}{j} = \frac{\alpha!}{j!(\alpha-j)!}$ represents the coefficient of recursive function.

In fact, the RL-FDD is obtained on the basis of GL-FDD by simplifying the calculation process. RL-FDD can be expressed as follows:

$$^{R}D_{t}^{\alpha}f(t) = \frac{1}{\Gamma(n-\alpha)} (\frac{d}{dt})^{n} \int_{t_0}^{t} \frac{f(\tau)}{(t-\tau)^{1+\alpha-n}} d\tau, \quad n-1 < \alpha < n, n \in N, \tag{13}$$

where $^{R}D_{t}^{\alpha}$ represents the RL-FDD type; n is an integer; and $\Gamma(\cdot)$ is the Gamma function, a commonly used basic functions in fractional calculus, defined as follows:

$$\Gamma(z) = \int_{0}^{\infty} t^{z-1} e^{-t} dt, \quad z \in C. \tag{14}$$

The function $\Gamma(\cdot)$ has the following properties:

$$\Gamma(z+1) = z\Gamma(z), \quad z \in N. \tag{15}$$

And the Laplace transform will be established by:

$$L[\frac{t^{\alpha-1}}{\Gamma(\alpha)} H(x)] = \frac{1}{s^{\alpha}}, \tag{16}$$

where $H(x)$ denotes the unit step function, which implies that it only needs $x \geq 0$. The formula and its inverse transformation are often used in fractional calculus.

The Caputo-FDD is expressed as follows:

$$^{C}D_{t}^{\alpha}f(t) = \frac{1}{\Gamma(n-\alpha)} \int_{t_0}^{t} \frac{f^{(n)}(\tau)}{(t-\tau)^{1+\alpha-n}} d\tau, \quad n-1 < \alpha < n, n \in N, \tag{17}$$

where $^{C}D_{t}^{\alpha}$ represents the Caputo-FDD type.

The Laplace transform of the Caputo-FDD is expressed as follows:

$$L\{^{C}D_{t}^{\alpha}f(t)\} = s^{\alpha} F(s) - \sum_{k=0}^{n-1} s^{\alpha-k-1} f^{(k)}(0). \tag{18}$$

Thus, the Laplace transform of the Caputo-FDD under the zero initial conditions is:

$$L\{^{C}_{0}D_{t}^{\alpha}f(t)\} = s^{\alpha} F(s). \tag{19}$$

The derivation of constants in the Caputo-FDD is bounded, while the derivation of constants in the RL-FDD is unbounded. The RL-FDD needs to solve an initial value problem that it is theoretically feasible but lacks physical meaning. Therefore, the Caputo-FDD is more suitable for solving the initial value problem of fractional calculus. Thus, it was adopted in this paper.

3.2. Fractional-Order KiBaM

The internal electrochemical reaction of a power battery is extremely complex. The strong nonlinear characteristics of LIBs shows a fractional-order dynamic behavior [21–23]. Therefore, fractional calculus can be used to model a novel fractional-order KiBaM (FO-KiBaM) with a higher accuracy [26]. Fractional derivatives can be replaced to describe the battery capacity change process in (2).

$$\begin{cases} \frac{d^\alpha y_1}{dt^\alpha} = -i(t) + k\delta_h(t) = -i(t) + k\left(\frac{y_2(t)}{1-c} - \frac{y_1(t)}{c}\right), \\ \frac{d^\alpha y_2}{dt^\alpha} = -k\delta_h(t) = -k\left(\frac{y_2(t)}{1-c} - \frac{y_1(t)}{c}\right), \end{cases} \quad (20)$$

where α is the order of fractional derivative equations, and $0 < \alpha < 1$.

In the case when the initial time $t_0 = 0$, and the battery is discharged at a constant current I, the Laplace transform of (20) will be expressed as:

$$\begin{cases} s^\alpha Y_1(s) = -\frac{I}{s} + k\left(\frac{Y_2(s)}{1-c} - \frac{Y_1(s)}{c}\right), \\ s^\alpha Y_2(s) = -k\left(\frac{Y_2(s)}{1-c} - \frac{Y_1(s)}{c}\right). \end{cases} \quad (21)$$

Similarly, assuming that the initial state of y_1 and y_2 are the same as in (10), the following will be obtained:

$$\begin{cases} Y_1(s) = \left(\frac{1-c}{s^\alpha+k'} + \frac{c}{s^\alpha}\right)\left(y_1(0) - \frac{I}{s}\right) + \left(\frac{c}{s^\alpha} - \frac{c}{s^\alpha+k'}\right)y_2(0) = \frac{cC_0}{s^\alpha} - \frac{cI}{s^{\alpha+1}} - \frac{(1-c)I}{s(s^\alpha+k')}, \\ Y_2(s) = \left(\frac{1-c}{s^\alpha} - \frac{1-c}{s^\alpha+k'}\right)\left(y_1(0) - \frac{I}{s}\right) + \left(\frac{1-c}{s^\alpha} + \frac{c}{s^\alpha+k'}\right)y_2(0) = \frac{(1-c)C_0}{s^\alpha} - \frac{(1-c)I}{s^{\alpha+1}} + \frac{(1-c)I}{s(s^\alpha+k')}. \end{cases} \quad (22)$$

The above inverse Laplace transform of the fractional calculus transfer function can be obtained using the Mittag-Leffler function, a commonly used basic function in fractional calculus [27–30]. The Mittag-Leffler function has two different definition forms: the single-parameter form and the two-parameter form. The two-parameter form of the Mittag-Leffler function is defined as shown below:

$$E_{\alpha,\beta}(z) = \sum_{j=0}^{\infty} \frac{z^j}{\Gamma(\alpha j + \beta)}, \quad \alpha > 0, \beta > 0. \quad (23)$$

If $\beta = 1$, the single-parameter form of the Mittag-Leffler function will be obtained:

$$E_\alpha(z) = \sum_{j=0}^{\infty} \frac{z^j}{\Gamma(\alpha j + 1)}, \quad \alpha > 0. \quad (24)$$

The exponential function e^z is critical in calculus. Similarly, the Mittag-Leffler function equally plays an important role, and it appears frequently in solutions of fractional differential equations. Sometimes the Mittag-Leffler function with two parameters is also called a generalized exponential function. In fact, e^z can be seen as a special case of the Mittag-Leffler function, because they are equal if $\alpha = 1$.

$$E_1(z) = \sum_{j=0}^{\infty} \frac{z^j}{\Gamma(j+1)} = \sum_{j=0}^{\infty} \frac{z^j}{j!} = e^z. \quad (25)$$

To facilitate the inverse Laplace transform, we defined a new function as shown below:

$$\varepsilon(t, m, \alpha, \beta) = t^{\beta-1} E_{\alpha,\beta}(mt^\alpha). \quad (26)$$

Its Laplace transform can be obtained as follows:

$$L[\varepsilon(t, \pm m, \alpha, \beta)] = \frac{s^{\alpha-\beta}}{s^\alpha \mp m}. \tag{27}$$

According to the above properties of (22), it is clear that $\beta = \alpha + 1$ in the transfer function (27), and its inverse Laplace transform will be obtained as shown below:

$$\begin{cases} y_1(t) = cC_0 \frac{t^{\alpha-1}}{\Gamma(\alpha)} - cI\frac{t^\alpha}{\Gamma(\alpha+1)} - (1-c)It^\alpha E_{\alpha,\alpha+1}(-k't^\alpha), \\ y_2(t) = (1-c)C_0 \frac{t^{\alpha-1}}{\Gamma(\alpha)} - (1-c)I\frac{t^\alpha}{\Gamma(\alpha+1)} + (1-c)It^\alpha E_{\alpha,\alpha+1}(-k't^\alpha). \end{cases} \tag{28}$$

Similarly, the height difference of two wells can be obtained:

$$\delta_h(t) = \frac{y_2(t)}{1-c} - \frac{y_1(t)}{c} = \frac{I}{c}t^\alpha E_{\alpha,\alpha+1}(-k't^\alpha). \tag{29}$$

Therefore, the capacity of the battery with fractional calculus can be expressed as follows:

$$\begin{cases} C_{rem}(t) = C_0 - It, \\ C_{unav}(t) = (1-c)\delta_h(t) = (1-c)\frac{I}{c}t^\alpha E_{\alpha,\alpha+1}(-k't^\alpha), \\ C_{av}(t) = C_0 - It - (1-c)\frac{I}{c}t^\alpha E_{\alpha,\alpha+1}(-k't^\alpha). \end{cases} \tag{30}$$

The SOC of the battery is similar to the fuel gauge of conventional internal combustion vehicles, which can be used to estimate the distance the vehicle can travel. The definition of SOC is:

$$SOC(t) = \frac{C_{rem}(t)}{C_{max}} \cdot 100\% = \frac{C_{t0} - \int_{t0}^{t} i(t)dt}{C_{max}} \cdot 100\%, \tag{31}$$

where C_{rem}, C_{max} represents the remaining capacity and the maximum available capacity of battery.

The unavailable capacity of the battery is not considered in this definition. Thus, it cannot tell the driver the actual available battery capacity at different C-rates. To predict the remaining mileage of electric vehicles more accurately, it is necessary to improve the SOC definition. An improved SOC definition is shown as below:

$$SOC(t) = \frac{C_{av}(t)}{C_{max}} \cdot 100\% = \frac{C_{t0} - \int_{t0}^{t} i(t)dt - C_{unav}(t)}{C_{max}} \cdot 100\%. \tag{32}$$

This definition of SOC takes into account the unavailable capacity of battery. Thus, it is useful in estimating the run time of the battery, and it can predict the remaining mileage of electric vehicles more accurately to relieve the "range anxiety" for the drivers. At the same time, by a more accurate definition of SOC, it can help to determine effective management strategies to avoid overcharging and over discharging the battery.

Further, we found that the height of y_1 in the proposed FO-KiBaM can be expressed as follows:

$$h_1(t) = \frac{y_1(t)}{c} = C_0 \left(\frac{t^{\alpha-1}}{\Gamma(\alpha)} - 1 \right) - I \left(\frac{t^\alpha}{\Gamma(\alpha+1)} - t \right) + C_{av}(t). \tag{33}$$

As can be seen from (32) and (33), the height h_1 of the left "well" reflects the change in the battery's remaining available capacity, and it also explains why the height h_1 is an intuitive representation of the battery SOC.

4. Parameter Identification and Experiment Verification

4.1. Experimental Platform and Test Results

As shown in Figure 4, the platform of the battery test system consisted of a battery charging and discharging cycler MKLtech MCT8-100-05Z, a programmable temperature chamber, a LiNiMnCoO ternary LIB module, and a computer with control software. The voltage and the current range of the battery cycler was 0–5 V and ±100 A, respectively. The temperature of the chamber was maintained at 30 °C. The sampling rate of voltage and current was set at 1 Hz. The tested ternary LIB module had a capacity of 32.50 Ah, which consisted of 13 battery cells connected in parallel, as shown in Figure 5. Table 1 shows the specifications of the ternary LIB cells. The available capacity tests at different C-rates of 0.2, 0.67, 1, 1.5, 2, and 3 C were carried out. Here, 1 C indicated that the rechargeable battery was continuously discharged for 1 h at a current equal to the battery's nominal capacity.

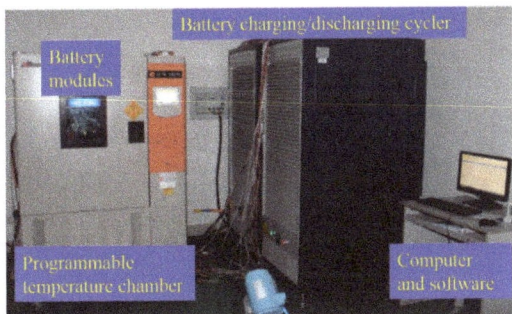

Figure 4. The battery test experimental platform.

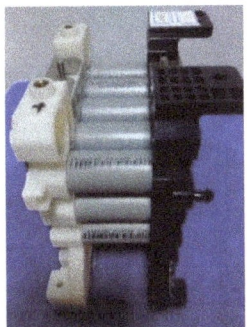

Figure 5. Battery module with 13 cells connected in parallel.

Table 1. Specification of the ternary lithium-ion battery (LIB) cell.

Parameter	Value	Parameter	Value
Type	LR1865SZ	Series-parallel	1S-13P
Nominal voltage	3.6 V	Rated capacity	2.5 Ah
End-of-charge voltage	4.2 V	End-of-discharge voltage	3.0 V

In the test, first the ternary battery module was fully charged by the constant current, constant voltage (CCCV) charging strategy. The constant current (CC) was 10 A, and the constant voltage (CV) was 4.2 V. The cutoff point for the CV charging stage was when the charging current was less than 1/50 C, or the CV charging time had reached 1 h. Then, the battery was left standing in the open-circuit state. In the discharging procedure, the battery was fully discharged to the cutoff voltage

(3.0 V) at 0.2 C. Then, the above test process was repeated to complete the available capacity tests at different C-rates. Throughout the testing, the discharge time, voltage, current, released capacity, and energy were monitored and recorded simultaneously. The current of the battery module in the charge and discharge processes at 2 C is shown in Figure 6.

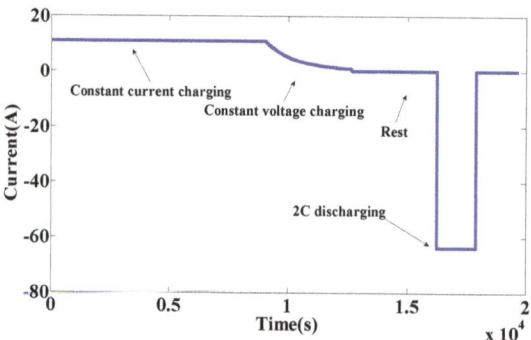

Figure 6. The battery testing currents.

The available capacity of the tested LIB module at different C-rates is shown in Table 2. It can be seen that the battery released 96.12% of the maximum capacity at 0.2 C; while the battery released 94.52% of the maximum capacity at 1 C. When the current was further increased to 3 C, the battery only released 84.89%. The result was consistent with the "capacity nonlinear effect", and the relationship of the available capacity and current was nonlinear.

Table 2. Available capacity and discharge time at different C-rates.

Discharge Rate (C)	Discharge Current (A)	Discharge Time (min)	Released Capacity (Ah)	Proportion (%)
0.2	6.410	292.5	31.24	96.12
0.67	21.26	87.34	30.95	95.23
1	31.88	57.81	30.72	94.52
1.5	47.83	37.56	29.94	92.12
2	63.78	27.39	29.11	89.57
3	95.69	17.30	27.59	84.89

4.2. Parameter Identification

The identification parameters of KiBaM, including capacity distribution ratio c, rate coefficient k, and fractional order α, were the key in achieving satisfactory accuracy based on the experimental data [31–33]. The identification parameters were mainly arranged as follows.

Distribution ratio (c): The battery was charged until full, SOC = 1, as the initial state; and then the battery was discharged at a large constant current, here it was 95.69 A (3 C). The released capacity was denoted as C_1, and as shown in Table 2, C_1 = 27.59 Ah. The maximum available capacity of the battery was 32.50 Ah, denoted as C_{max}; then, $c = C_1/C_{max} = 0.849$.

Rate coefficient (k): The available capacity of the battery was tested at different C-rates. For the classic KiBaM, we only needed one set of experimental data, and k' could be obtained. The data measured at 1 C were selected here, and the data at other C-rates were used to verify the model's accuracy later.

$$(32.5 - 30.72) \cdot 3600 = (1 - 0.849)\frac{31.88}{0.849}\frac{1 - e^{-3468.6k'}}{k'}. \tag{34}$$

A new function $F(k')$ can be defined and modified as follows:

$$F(k') = k' - (1 - 0.849)\frac{31.88}{0.849}\frac{1 - e^{-3468.6k'}}{1.78\cdot3600}. \tag{35}$$

The change of the function $F(k')$ with k' is depicted in Figure 7, and the solution, $k' = 0.000836$, could be confirmed when $F(k')$ is 0. Thus, the unavailable capacity of the tested battery in KiBaM was obtained as follows:

$$C_{\text{unav}}(t) = (1 - 0.849)\frac{I}{0.849}\frac{1 - e^{-0.000836t}}{0.000836}. \tag{36}$$

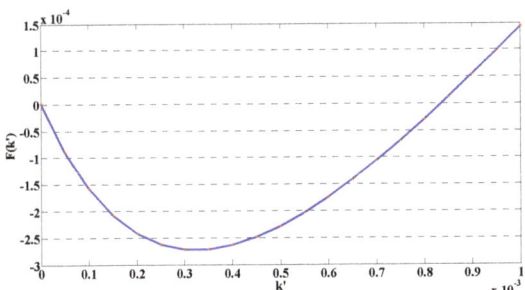

Figure 7. The function curve that changes as k' changes.

As for the FO-KiBaM, k' and α of the tested battery module could also be identified by using one set of experimental data at 1 C.

$$(32.5 - 30.72)\cdot 3600 = (1 - 0.849)\frac{31.88}{0.849}3468.6^\alpha E_{\alpha,\alpha+1}(-k'3468.6^\alpha). \tag{37}$$

And a new function $F(\alpha, k')$ could be defined and modified as follows:

$$F(\alpha, k') = 1.78\cdot 3600 - (1 - 0.849)\frac{31.88}{0.849}3468.6^\alpha E_{\alpha,\alpha+1}(-k'3468.6^\alpha). \tag{38}$$

The change of the function $F(\alpha, k')$ with k' and α is shown in Figure 8, where α changed from 0.1 to 0.9. We can see that it tended towards 0 when α was 0.9. Further, its trend was depicted in Figure 9 when α changed from 0.91 to 0.99. It was revealed that the numerical solution of $F(\alpha, k')$ was not unique. The values of k' and α could be a combination within a suitable range. Here, the parameters $k' = 0.000689$ and $\alpha = 0.99$ were selected for the FO-KiBaM. Thus, the unavailable capacity of the tested battery in FO-KiBaM was obtained.

$$C_{\text{unav}}(t) = (1 - 0.849)\frac{I}{0.849}t^{0.99}E_{0.99,1.99}(-0.000689t^{0.99}). \tag{39}$$

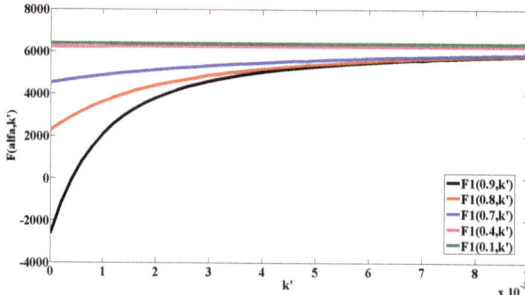

Figure 8. Trend of the function as k' and α (0.1 to 0.9) changes.

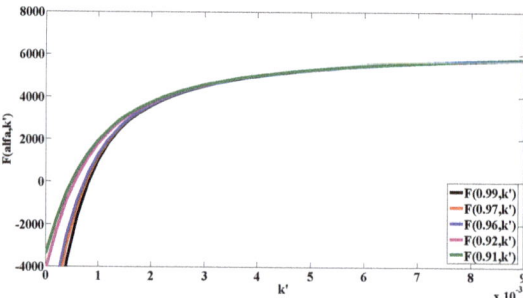

Figure 9. Trend of the function as k' and α (0.91 to 0.99) changes.

4.3. Experiment Verification

The effect of the C-rate on the available capacity of the battery was verified by experiment. The variation of the total capacity and unavailable capacity of the battery at different C-rates are depicted in Figures 10 and 11, in which the discharge was finished when the total capacity of the battery (black line) intercepted the unavailable capacity (red/blue lines). At this point, all the remaining capacity of the battery was the unavailable capacity, so the discharge process was over. The prediction results of the available capacity of the battery are listed in Table 3. It can be seen that compared with the classic KiBaM, the errors of available capacity in the proposed FO-KiBaM were less, and it performed well over a wid applied current range. Specifically, its mean absolute error (MAE) was only 1.91%, with an improvement of 0.44%. Although the accuracy of KiBaM was already high, the proposed FO-KiBaM still had a smaller fitting error over a wide applied current range. Further study on the state estimations of LIBs and the high-precision optimization control of energy management will be useful, especially with the FO-KiBaM.

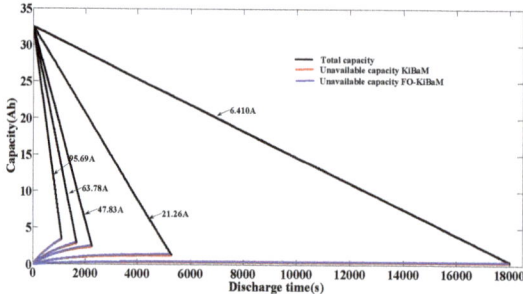

Figure 10. Battery capacity variance at different C-rates.

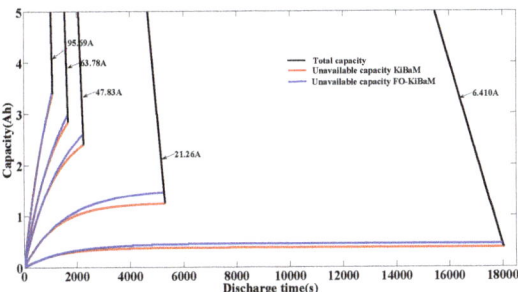

Figure 11. Battery capacity variance at different C-rates (local magnification).

Table 3. Prediction results of the battery's available capacity at different C-rates.

Discharge Current (A)	Available Capacity (Ah)			Improve Accuracy (%)
	Experimental Data	KiBaM	FO-KiBaM	
6.410	31.24	32.12	32.04	0.26
21.26	30.95	31.27	31.03	0.78
47.83	29.94	30.10	29.90	0.67
63.78	29.11	29.66	29.50	0.55
95.69	27.59	29.11	29.04	0.25

It should be pointed out that if the estimation accuracy was higher in some applications, we could achieve parameter identification according to current interval (segmentation) by using several sets of the tested data instead of just one set, so that the accuracy of the model could be improved. If we thought the parameters of the model were related to the current, we could even identify the parameters under different currents. Thus, that the accuracy of the model can be further improved, but the amount of calculation would also increase. The whole process of parameter identification is the same, whether it is how to select the data or how many sets of data will be weighed. However, if the accuracy of the model identified by only one set of data met the application requirements, it would significantly reduce the amount of calculation. For a fair comparison and reduction in the number of calculations, we only selected one set of data to identify the parameters of both the KiBaM and the FO-KiBaM.

5. Conclusions

In recent years, various families of fractional-order systems have been found to be remarkably important and fruitful. Fractional calculus plays an important role in complex systems and, therefore, allows us to better describe real-world phenomena. By obtaining more parameters and degrees of freedom, fractional-order models can describe nonlinear performances of complex systems more accurately. Due to specific material and chemical properties of batteries, fractional calculus is more reasonable to describe the nonlinear performance of a battery's capacity. The proposed FO-KiBaM can describe the battery's nonlinear characteristics more accurately, with greater flexibility and novelty compared to the classic KiBaM. The proposed model can be applied in engineering. The estimation of the available capacity of LIBs is meaningful with a wide applicable current range, and only a set of data at one C-rate is needed to accurately estimate the available capacity at different rates, which greatly reduces the number of calculations. What is more, the proposed FO-KiBaM provides a basic battery model for further research on SOC estimation methods.

6. Patents

A patent, termed fractional order KiBaM (kinetic battery model) that considers nonlinear capacity characteristics and parameter identification methods, resulted from the work reported in this manuscript, which can be seen on academic websites such as Google Patents. The progress status of this patent is as follows. Application filed by Shandong University and priority to CN201710093350.0A on 21 February 2017, publication of CN106855612A on 16 June 2017, and Notice of First Review on 16 November 2018. PCT/CN2017/106912 application filed by Shandong University on 19 October 2017, and publication of WO2018153116A1 on 30 August 2018.

Author Contributions: Conceptualization, Q.Z. and Y.S.; methodology, Q.Z. and Y.S.; software, Q.Z. and Y.L.; validation, Q.Z.; formal analysis, Q.Z. and Y.L.; writing—original draft preparation, Q.Z.; writing—review and editing, N.C. and B.D.; project administration, N.C., B.D., and C.Z.; funding acquisition, N.C., B.D., and C.Z.

Funding: This research was funded by the National Natural Science Foundation of China (Grant No.61527809, U1864205, U1764258, and 61633015) in part, and the National Key Research and Design program of China (Grant No. 2018YFB0104000) in part, which are gratefully acknowledged. Most importantly, the authors would also like to thank the anonymous reviewers for their valuable comments and suggestions.

Acknowledgments: The English writing of this manuscript has been carefully edited by a native English speaker. The authors would like to thank him for the proofreading and editing. Special thanks to Professor Igor Podlubny for sharing the Matlab routine for evaluating the Mittag-Leffler function with two parameters in Matlab Central at https://www.mathworks.com/matlabcentral/fileexchange/8738-mittag-leffler-function.

Conflicts of Interest: The authors declare no conflict of interest.

Abbreviations

LIB	lithium-ion battery
EchM	electrochemical model
AM	analytical model
SM	stochastic model
NNM	neural network model
ECM	equivalent circuit model
U-I	voltage-current
KiBaM	kinetic battery model
T-KiBaM	temperature-dependent kinetic battery model
FO-KiBaM	fractional-order kinetic battery model
BMS	battery management system
Ah	ampere hours
SOC	state of charge
SOH	state of health
SOP	state of peak power
SOF	state of function
SOE	state of energy
C-rate	current rate
CCCV	constant-current and constant-voltage
CC	constant current
CV	constant voltage
FDD	fractional derivative definition
GL-FDD	Grunwald-Letnikov fractional derivative definitions
RL-FDD	Riemann-Liouville fractional derivative definitions
MAE	mean absolute error

Nomenclature

y_1	directly available capacity
h_1	height of directly available capacity
y_2	temporary capacity
h_2	height of temporary capacity well
δ_h	the height difference of two wells
k	rate that the charge flows from y_2 into y_1
c	capacity proportion of two wells
i	discharge current
I	the constant discharge current
t_0	the initial time
t_d	the discharge end time
t_r	the recovery end time
C_{av}	available capacity of battery
C_{unav}	unavailable capacity of battery
C_{rem}	remaining available capacity of battery
C_{max}	the maximum available capacity of battery
C_{t0}	initial capacity of battery

References

1. Lievre, A.; Sari, A.; Venet, P.; Hijazi, A.; Ouattarabrigaudet, M.; Pelissier, S. Practical online estimation of lithium-ion battery apparent series resistance for mild hybrid vehicles. *IEEE Trans. Veh. Technol.* **2016**, *65*, 4505–4511. [CrossRef]
2. Xia, B.; Nguyen, T.; Mi, C. A correlation-based fault detection method for short circuits in battery packs. *J. Power Sources* **2017**, *337*, 1–10. [CrossRef]
3. Shang, Y.; Zhang, Q.; Cui, N.; Zhang, C. A cell-to-cell equalizer based on three-resonant-state switched-capacitor converters for series-connected battery strings. *Energies* **2017**, *10*, 206. [CrossRef]
4. Hannan, M.A.; Lipu, M.S.H.; Hussain, A.; Mohamed, A. A review of lithium-ion battery state of charge estimation and management system in electric vehicle applications: Challenges and recommendations. *Renew. Sustain. Energy Rev.* **2017**, *78*, 834–854. [CrossRef]
5. Rivera-Barrera, J.P.; Muñoz-Galeano, N.; Sarmiento-Maldonado, H.O. SoC Estimation for Lithium-ion Batteries: Review and Future Challenges. *Electronics* **2017**, *6*, 102. [CrossRef]
6. Lai, X.; Yi, W.; Zheng, Y.; Zhou, L. An All-Region State-of-Charge Estimator Based on Global Particle Swarm Optimization and Improved Extended Kalman Filter for Lithium-Ion Batteries. *Electronics* **2018**, *7*, 321. [CrossRef]
7. Jongerden, M.R.; Haverkort, B.R. Which battery model to use? *IET Softw.* **2009**, *3*, 445–457. [CrossRef]
8. Ko, S.T.; Ahn, J.H.; Lee, B.K. Enhanced Equivalent Circuit Modeling for Li-ion Battery Using Recursive Parameter Correction. *J. Electr. Eng. Technol.* **2018**, *13*, 1147–1155.
9. He, H.; Xiong, R.; Guo, H.; Li, S. Comparison study on the battery models used for the energy management of batteries in electric vehicles. *Energy Convers. Manag.* **2012**, *64*, 113–121. [CrossRef]
10. Shang, Y.; Zhang, Q.; Zhang, C.; Cui, N. Research on variable-order RC equivalent circuit model for lithium-ion battery based on the AIC criterion. *Trans. China Electrotech. Soc.* **2015**, *30*, 55–62.
11. Zhou, Y.; Huang, M. On-board Capacity Estimation of Lithium-ion Batteries Based on Charge Phase. *J. Electr. Eng. Technol.* **2018**, *13*, 733–741.
12. Cloth, L.; Jongerden, M.R.; Haverkort, B.R. Computing battery lifetime distributions. In Proceedings of the 37th Annual IEEE/IFIP International Conference on Dependable Systems and Networks (DSN'07), Edinburgh, UK, 25–28 June 2007; pp. 780–789.
13. Manwell, J.; McGowan, J. Lead acid battery storage model for hybrid energy systems. *Sol. Energy* **1993**, *50*, 399–405. [CrossRef]
14. Kim, T.; Qiao, W. A hybrid battery model capable of capturing dynamic circuit characteristics and nonlinear capacity effects. *IEEE Trans. Energy Convers.* **2011**, *26*, 1172–1180. [CrossRef]
15. Blanco, C.; Sánchez, L.; González, M.; Antón, J.C.; García, V.; Viera, J.C. An equivalent circuit model with variable effective capacity for lifepo4 batteries. *IEEE Trans. Veh. Technol.* **2014**, *63*, 3592–3599. [CrossRef]

16. Sánchez, L.; Blanco, C.; Antón, J.C.; García, V.; González, M.; Viera, J.C. A variable effective capacity model for LiFePO4 traction batteries using computational intelligence techniques. *IEEE Trans. Ind. Electron.* **2015**, *62*, 555–563. [CrossRef]
17. Baert, D.; Vervaet, A. Lead-acid battery model for the derivation of Peukert's law. *Electrochim. Acta* **1999**, *44*, 3491–3504. [CrossRef]
18. Rodrigues, L.M.; Montez, C.; Moraes, R.; Portugal, P.; Vasques, F. A Temperature-Dependent Battery Model for Wireless Sensor Networks. *Sensors* **2017**, *17*, 422. [CrossRef]
19. Freeborn, T.J.; Maundy, B.; Elwakil, A.S. Fractional-order models of supercapacitors, batteries and fuel cells: A survey. *Mater. Renew. Sustain. Energy* **2015**, *4*, 9. [CrossRef]
20. Wang, W.G.; Li, Y.; Chen, Y.Q. Ubiquitous fractional order capacitors. In Proceedings of the International Conference on Fractional Dierentiation and Its Applications, Novi Sad, Serbia, 18–20 July 2016.
21. Malek, H.; Dadras, S.; Chen, Y. Fractional order equivalent series resistance modelling of electrolytic capacitor and fractional order failure prediction with application to predictive maintenance. *IET Power Electron.* **2016**, *9*, 1608–1613. [CrossRef]
22. Sabatier, J.; Merveillaut, M.; Francisco, J.; Guillemard, F.; Porcelatto, D. Fractional models for lithium-ion batteries. In Proceedings of the 2013 European Control Conference, Zürich, Switzerland, 17–19 July 2013; pp. 3458–3463.
23. Sabatier, J.; Cugnet, M.; Laruelle, S.; Grugeon, S.; Sahut, B.; Oustaloupa, A.; Tarasconb, J.M. A fractional order model for lead-acid battery crankability estimation. *Commun. Nonlinear Sci. Numer. Simul.* **2010**, *15*, 1308–1317. [CrossRef]
24. Sun, Y.; Wu, X.; Cao, J.; Wei, Z.; Sun, G. Fractional extended Kalman filtering for nonlinear fractional system with Lévy noises. *IET Control Theory Appl.* **2017**, *11*, 349–358. [CrossRef]
25. Sun, H.; Zhang, Y.; Wei, S.; Zhu, J.; Chen, W. A space fractional constitutive equation model for non-Newtonian fluid flow. *Commun. Nonlinear Sci. Numer. Simul.* **2018**, *62*, 409–417. [CrossRef]
26. Zhang, Q.; Shang, Y.; Cui, N.; Li, Y.; Zhang, C. A fractional-order KiBaM of lithium-ion batteries with capacity nonlinearity. In Proceedings of the 2017 Chinese Automation Congress, Jinan, China, 20–22 October 2017; pp. 4995–5000.
27. Li, Y.; Chen, Y.Q.; Podlubny, I. Mittag-Leffler stability of fractional order nonlinear dynamic systems. *Automatica* **2009**, *45*, 1965–1969. [CrossRef]
28. Li, Y.; Chen, Y.Q.; Podlubny, I. Stability of fractional-order nonlinear dynamic systems: Lyapunov direct method and generalized Mittag-Leffler stability. *Comput. Math. Appl.* **2010**, *59*, 1810–1821. [CrossRef]
29. Wang, Q.; Zhang, J.; Ding, D.; Qi, D. Adaptive Mittag-Leffler stabilization of a class of fractional order uncertain nonlinear systems. *Asian J. Control* **2016**, *18*, 2343–2351. [CrossRef]
30. Khurram, A.; Rehman, H.; Mukhopadhyay, S.; Ali, D. Comparative Analysis of Integer-order and Fractional-order Proportional Integral Speed Controllers for Induction Motor Drive Systems. *J. Power Electron.* **2018**, *18*, 723–735.
31. Yang, J.; Xia, B.; Shang, Y.L.; Huang, W.; Mi, C. Improved battery parameter estimation method considering operating scenarios for HEV/EV applications. *Energies* **2016**, *10*, 5. [CrossRef]
32. Feng, F.; Lu, R.; Wei, G.; Zhu, C. Identification and analysis of model parameters used for LiFePO$_4$ cells series battery pack at various ambient temperature. *IET Electr. Syst. Transp.* **2016**, *6*, 50–55. [CrossRef]
33. Hua, C.C.; Fang, Y.H.; Chen, Y.L. Modified rectifications for improving the charge equalisation performance of series-connected battery stack. *IET Power Electron.* **2016**, *6*, 1924–1932. [CrossRef]

© 2019 by the authors. Licensee MDPI, Basel, Switzerland. This article is an open access article distributed under the terms and conditions of the Creative Commons Attribution (CC BY) license (http://creativecommons.org/licenses/by/4.0/).

Article

State of Charge Estimation for Lithium-Ion Batteries Based on Temperature-Dependent Second-Order RC Model

Yidan Xu [1], Minghui Hu [1,*], Chunyun Fu [1,*], Kaibin Cao [1], Zhong Su [2] and Zhong Yang [2]

1. State Key Laboratory of Mechanical Transmissions, School of Automotive Engineering, Chongqing University, Chongqing 400044, China
2. Chongqing Changan Automobile Co., Ltd., Chongqing 400023, China
* Correspondence: hu_ming@cqu.edu.cn (M.H.); fuchunyun@cqu.edu.cn (C.F.)

Received: 19 August 2019; Accepted: 7 September 2019; Published: 10 September 2019

Abstract: Accurate estimation of battery state of charge (SOC) is of great significance for extending battery life, improving battery utilization, and ensuring battery safety. Aiming to improve the accuracy of SOC estimation, in this paper, a temperature-dependent second-order RC equivalent circuit model is established for lithium-ion batteries, based on the battery electrical characteristics at different ambient temperatures. Then, a dual Kalman filter algorithm is proposed to estimate the battery SOC, using the proposed equivalent circuit model. The SOC estimation results are compared with the SOC value obtained from experiments, and the estimation errors under different temperature conditions are found to be within ±0.4%. These results prove that the proposed SOC estimation algorithm, based on a temperature-dependent second-order RC equivalent circuit model, provides accurate SOC estimation performance with high temperature adaptability and robustness.

Keywords: lithium-ion battery; temperature-dependent second-order RC model; SOC estimation; dual Kalman filter

1. Introduction

One fundamental challenge in the commercialization of electric vehicles is the battery system, and a safe and efficient battery system hinges on a reliable battery management system (BMS) [1]. At present, one of the main difficulties that hinder the development of BMS technology is the accurate estimation of state of charge (SOC). It has been pointed out that accurate SOC information is conducive to protecting batteries, preventing overcharging and over-discharging, improving battery utilization, and increasing the cruising range of electric vehicles [2–4].

However, the SOC cannot be directly monitored because of the battery systems' nonlinearity, time-varying characteristics, and the complexity of electrochemical reactions [5]. To tackle this problem, plenty of SOC estimation methods have been proposed in the literature. The existing SOC estimation methods can be divided into two major categories: non-model-based methods and model-based methods. The non-model-based methods mainly include the following: ampere hour (Ah) integration method [6], open circuit voltage (OCV) method [7], internal resistance method [8], and machine learning algorithms [9]. The model-based methods mainly include the following: particle filter (PF) [10], Kalman filter (KF) [11], and its improved algorithms [12–14].

Among the non-model-based methods, the Ah integration method is the simplest and most commonly used. Its implementation is straightforward and its computation load is low [15]. However, as an open-loop algorithm, the existence of uncertainties, such as noise, temperature, and current variations, can lead to large errors. The OCV method and internal resistance method rely on the correlation between the battery SOC and its external static characteristic parameter (i.e., OCV or internal resistance), and employ a look-up table to determine the estimated SOC value. However, the measurement of battery OCV requires even distribution of the electrolyte inside the battery. This process takes fairly long, which makes real-time measurement of OCV impossible [16]. Similarly, real-time measurement of battery internal resistance is also very difficult. As a result, neither the OCV method nor the internal resistance method can be used independently in practical applications. Besides, the machine learning algorithms are devised based on various mechanisms such as artificial neural networks [17], fuzzy logic inference [18], and support vector regression (SVR) [19]. These algorithms require a large amount of training data to establish the nonlinear relationship between the input to the battery and the output from the battery [20,21]. The performance of these algorithms is highly dependent on the quantity and quality of the training data, which in turn restricts the applicability and accuracy of these methods.

The model-based methods are more practical in terms of SOC online estimation. These methods rely on a high-precision battery model and adopt a closed-loop structure to perform SOC estimation with an unknown initial SOC value. These methods iteratively correct the SOC estimation error, using the difference between the measured terminal voltage and the battery model output. The PF [22] can achieve good results with non-Gaussian white noise; however, it leads to higher computational load compared with the KF. The KF is widely used owing to its capability of finding the optimal solution of linear Gaussian systems. In the work of [23], the KF is employed in conjunction with the Ah integration method, which improves the SOC estimation accuracy for lithium-ion batteries. Various improved KF algorithms, such as extended Kalman filter (EKF) [11] and double extended Kalman filter [24], have been extensively studied in battery SOC estimation. In the work of [25], an EKF based on the second-order RC model is proposed to reduce the influence of noise during the measurement process.

The current literature mainly focuses on the introduction of SOC estimation methods and their advantages and disadvantages. The SOC estimation errors for the existing methods are normally investigated under some specific scenarios. Duong et al. [26] proposed a multiple adaptive forgetting factors recursive least-squares (MAFF-RLS) technique for LiFePO4 battery SOC estimation. The maximum SOC estimation error is found to be 2.8% in two standard driving cycles, Urban Dynamometer Driving Schedule (UDDS) and New European Driving Cycle (NEDC). Yang et al. [27] compared the performances of several algorithms (i.e., EKF, UKF, and PF) for cylindrical-type 18,650 (LiFePO4) battery SOC estimation under a combined dynamic loading profile, which is composed of the dynamic stress test, the federal urban driving schedule, and the US06. In most circumstances, the three algorithms provided satisfactory SOC estimation results, as well as small root mean square errors (RMSEs) (less than 4%).

In order to further improve the accuracy of SOC estimation, in this paper, the electrical characteristics of lithium-ion batteries under different ambient temperatures are analyzed, and a temperature-dependent second-order RC model is established. Then, a dual Kalman filter (DKF) algorithm is proposed based on the established model, which synthesizes the Ah integration method, the KF algorithm, and the EKF algorithm. The accuracy and temperature adaptability of the proposed SOC estimation algorithm are verified through experiments. The proposed battery model and SOC estimation method have taken into consideration the important effects of ambient temperature, and the results of this study can assist with the improvement of battery thermal management systems and enhance the reliability of electric vehicles operating in all climate conditions.

The rest of the paper is organized as follows. Section 2 describes the establishment of the temperature-dependent second-order RC model after analyzing the characteristics of lithium-ion batteries, Section 3 introduces the identification of unknown parameters involved in this model, Section 4 elaborates on the model verification process under different temperature conditions, Section 5 proposes a DKF algorithm for SOC estimation and its verification, and Section 6 concludes the paper.

2. Temperature-Dependent Second-Order RC Model

The battery characteristics vary greatly at different temperatures. In order to improve the temperature adaptability of the battery model, temperature variation factors should be taken into account in the battery modeling process. In this section, the electrical characteristics of lithium-ion batteries at different ambient temperatures are firstly introduced. Then, a temperature-dependent second-order RC model is established, taking into account the effects of ambient temperature variations.

2.1. Characteristics of Lithium-Ion Batteries

When the current of the battery external circuit is zero and the internal electrochemical reactions are in equilibrium, the potential difference between the positive and negative electrodes is called the OCV [28]. The OCV is closely related to the battery SOC and the ambient temperature, and it is crucial for lithium-ion battery modeling and SOC estimation. As shown in Figure 1a, the OCV of the lithium-ion battery gradually decreases as the ambient temperature rises.

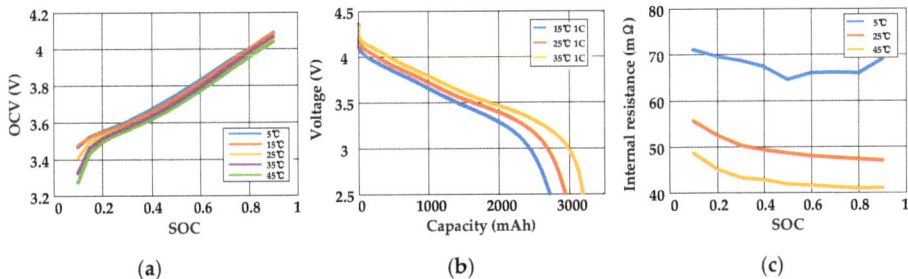

Figure 1. Battery characteristics under different ambient temperature conditions. (**a**) Relationship between state of charge (SOC) and open circuit voltage (OCV), (**b**) battery maximum discharge capacity, and (**c**) relationship between SOC and internal resistance.

With a certain discharge rate and a certain cutoff voltage, the amount of electricity that the battery can discharge is defined as the battery capacity [29]. The battery capacities at different ambient temperatures are shown in Figure 1b, where we see that the battery capacity increases as the ambient temperature increases.

Figure 1c shows how the battery internal resistance varies with the ambient temperature and SOC. We observe that the internal resistance decreases dramatically with the increase of ambient temperature; however, the influence of SOC on the internal resistance is insignificant.

2.2. Battery Modeling

On the basis of the influence of ambient temperature on lithium-ion batteries introduced in Section 2.1, a temperature-dependent second-order RC equivalent circuit model is established in this section, taking into account both model accuracy and model complexity. This model is composed of three modules: the OCV module, the internal resistance module, and the RC network module. The structure of the proposed model is shown in Figure 2, where V_t represents the battery terminal voltage, V_{ocv} indicates the OCV, V_1 and V_2 denote the voltages generated by the polarization phenomenon, I stands for the current (positive for charging and negative for discharging), T represents the ambient temperature, R_0 is the ohmic internal resistance, R_1 and R_2 are the polarization internal resistances, and C_1 and C_2 are the polarization capacitances. It must be pointed out that the effects of SOC, temperature, and current direction changes on the above parameters have been taken into account in the battery modeling, as we will explain in Section 3.

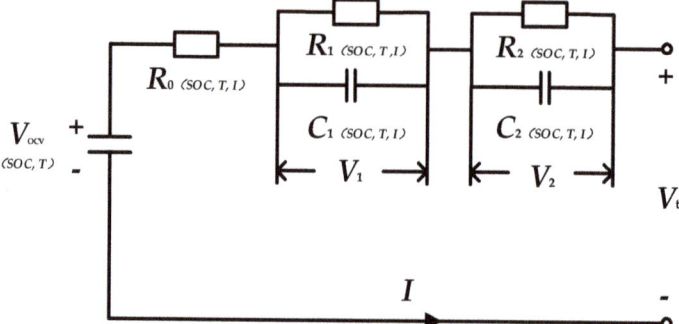

Figure 2. Structure of the proposed temperature-dependent second-order RC equivalent circuit model.

According to Kirchhoff's laws of voltage and current, the polarization voltages V_1 and V_2 satisfy the following rules:

$$\dot{V}_1 = -\frac{V_1}{R_1 C_1} + \frac{I}{C_1}, \tag{1}$$

$$\dot{V}_2 = -\frac{V_2}{R_2 C_2} + \frac{I}{C_2}, \tag{2}$$

where $\dot{V}_1$ and $\dot{V}_2$ denote the voltage change rates. The terminal voltage V is given by the following:

$$V = V_{\text{OCV}} + R_0 I + V_1 + V_2. \tag{3}$$

The battery SOC is dependent on the available capacity and current direction, and it can be expressed as follows:

$$SOC(t) = SOC(t_0) + \int_{t_0}^{t} \frac{\eta I(\tau)}{Q} d\tau, \tag{4}$$

where $SOC(t)$ and $SOC(t_0)$ are the SOC values at time t and t_0, respectively; η denotes the charging/discharging efficiency (0.98 for charging and 1 for discharging); and Q represents the maximum available capacity. Assuming that the sampling time is ΔT, discretizing Equations (1), (2) and (4) yields the following:

$$V_1(k) = exp(\frac{-\Delta T}{R_1 C_1}) V_1(k-1) + R_1 I(k)[1 - exp(\frac{-\Delta T}{R_1 C_1})], \tag{5}$$

$$V_2(k) = exp(\frac{-\Delta T}{R_2 C_2}) V_2(k-1) + R_2 I(k)[1 - exp(\frac{-\Delta T}{R_2 C_2})], \tag{6}$$

$$SOC(k) = SOC(k-1) + \frac{\eta t}{Q} I(k). \tag{7}$$

Equations (3) and (5)–(7) constitute the mathematical representation of the proposed temperature-dependent second-order RC model for lithium-ion batteries. These equations describe the dynamic characteristics of lithium-ion batteries at different temperatures, in a simple mathematical form with limited number of parameters.

Similar to our previous work [30], some parameters in this proposed battery model are not known a priori and need to be determined for model implementation. The parameters to be identified are R_1, R_2, C_1, C_2, R_0, V_{OCV}, and Q. In the following section, we shall explain in detail how these unknown parameters can be identified.

3. Model Parameter Identification

As mentioned above, the proposed battery model involves parameters that cannot be directly measured. Hence, in this section, we explain the detailed process for identifying these unknown parameters by means of experiments.

3.1. Experiment Specifications

The equipment used in our experimentation includes an incubator (HL404C, Well Test Equipment Co., Ltd., Chongqing, China), a battery testing device (BTS-5V100A, Neware Co., Ltd., Shenzhen, China), a set of measurement and control software, and a computer. In this study, a time interval of 0.1 s is used for data acquisition. The complete battery test system is shown in Figure 3. The Panasonic 18,650 ternary lithium-ion battery is used for testing, and its specifications are shown in Table 1.

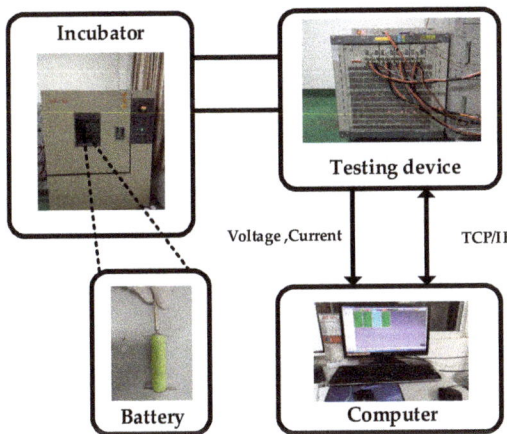

Figure 3. Battery test system.

Table 1. Specifications of the Panasonic 18,650 ternary lithium-ion battery.

Battery Model	Nominal Capacity	Rated Voltage	Charge Cutoff Voltage	Discharge Cutoff Voltage
18,650	3350 mAh	3.6 V	4.2 V	2.5 V

In order to identify the model parameters, the static capacity test (SCT), hybrid pulse power characteristic (HPPC) test, and double pulse discharge test were performed in this study, at five different ambient temperatures (5 °C, 15 °C, 25 °C, 35 °C, and 45 °C), according to the work of [31]. The SCT was used for battery capacity (Q) identification, and the double pulse discharge test was utilized to obtain the OCV (V_{OCV}) under specific SOC values. Parameters such as ohmic internal resistance (R_0), polarization resistances (R_1, R_2), and polarization capacitances (C_1, C_2) were obtained by the HPPC test. The major test steps of the three tests are shown in Table 2.

Table 2. Major test steps of the static capacity test (SCT), hybrid pulse power characteristic (HPPC), and double pulse tests.

Test Name		Test Steps
SCT Test	1	Battery shelved for 1 h
	2	0.2 C constant current charging to charge cut-off voltage
	3	Constant voltage charging to charge cut-off current
	4	Battery shelved for 2 h
	5	0.2 C constant current discharging to discharge cut-off voltage
HPPC Test	1	1 C constant current charging to charge cut-off voltage
	2	Constant voltage charging to charge cut-off current
	3	1 C constant current constant capacity (0.1 SOC) discharging
	4	Battery shelved for 1 h
	5	1 C constant current discharging for 10 s, battery shelved for 40 s, 0.75 C constant current charging for 10 s
	6	Repeat steps 3~5 to obtain battery internal resistances and power characteristics at different SOC values
Double Pulse Test	1	1 C constant current discharging to discharge cut-off voltage
	2	Battery shelved for 1 h
	3	0.5 C constant current constant capacity (0.1 SOC) charging
	4	Battery shelved for 2 min
	5	Repeat steps 1 and 2 until the voltage reaches the charge cut-off voltage
	6	Battery shelved for 12 h
	7	0.5 C constant current constant capacity (0.1 SOC) discharging
	8	Battery shelved for 2 min
	9	Repeat steps 1 and 2 until the voltage reaches the discharge cut-off voltage

3.2. Identification of OCV

In practice, the battery OCV cannot be directly measured. However, there exists a certain correspondence between the OCV and SOC, which is crucial for OCV estimation. This relationship can be expressed by the following empirical equation [32]:

$$V_{OCV} = K_1 + K_2 SOC + K_3 SOC^2 + K_4 SOC^3 + K_5 SOC^4, \qquad (8)$$

where K_i (i = 1, 2, ... 5) are five coefficients depending on the ambient temperature. The values of K_i determine the accuracy of the empirical Equation (8), and they can be obtained from the double pulse discharge test. The obtained coefficients of Equation (8) at different ambient temperatures are shown in Table 3, and the resulting 3D SOC–Temperature–OCV (SOC–T–OCV) map is shown in Figure 4. This SOC–T–OCV map is indeed a 3D look-up table, from which we can determine the OCV for certain values of SOC and ambient temperature.

Table 3. Coefficient values at different ambient temperatures.

Coefficient	5 °C	15 °C	25 °C	35 °C
K_1	2.7758	2.81055	2.58785	2.4553
K_2	1.6295	1.48735	1.78815	1.96495
K_3	−0.04675	−0.04095	−0.07095	−0.1013
K_4	−0.43585	−0.4026	−0.5894	−0.7354
K_5	0.07365	0.02485	0.05735	0.05175

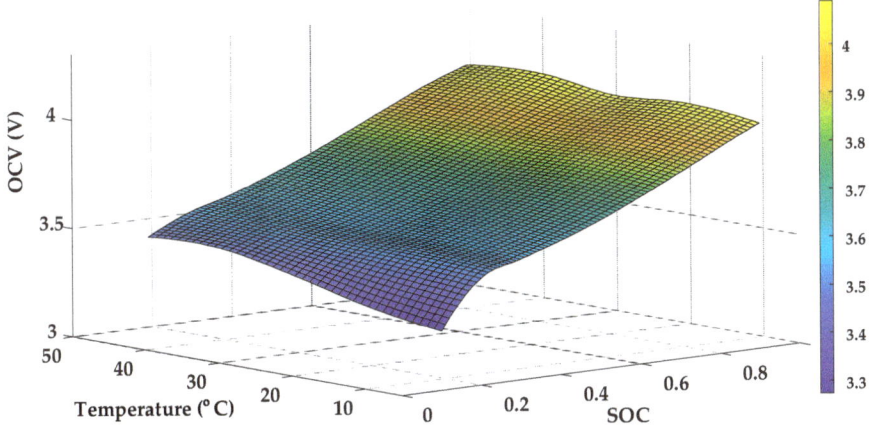

Figure 4. SOC–Temperature–OCV map.

3.3. Identification of Internal Resistances and Capacitances

In this section, we explain how to identify the internal resistances (R_0, R_1, and R_2) and capacitances (C_1 and C_2), by means of the HPPC test. In the HPPC test, the voltage and current vary according to the patterns shown in Figure 5. In this figure, one of the charge–discharge cycles is enlarged to clearly show the details.

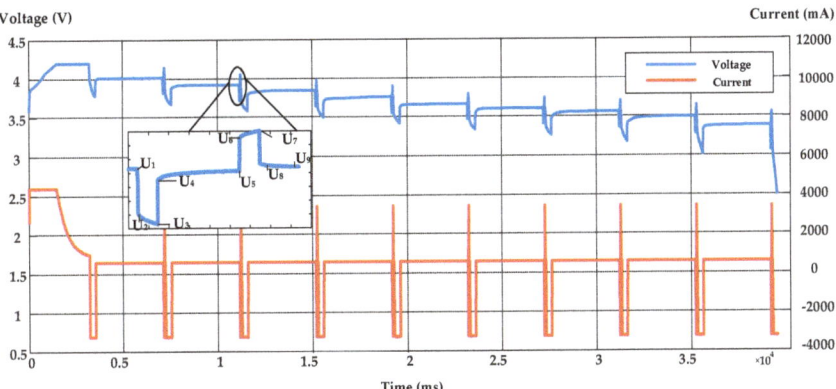

Figure 5. Voltage and current in the hybrid pulse power characteristic (HPPC) test.

Considering the discharge relaxation response curve ($U_4 \sim U_5$) and the charge relaxation response curve ($U_8 \sim U_9$) in Figure 5, the battery terminal voltage expression (i.e. Equation (3)) can be converted to the following equation, according to the work of [33]:

$$V(t) = a + b \times exp(-t/\tau_1) + c \times exp(-t/\tau_2), \tag{9}$$

where a, b, and c are the fitting coefficients; and τ_1 and τ_2 are the time constants of the two RCs in Figure 2. The five coefficients a, b, c, τ_1, and τ_2 in Equation (9) can be obtained by means of curve fitting using MATLAB software.

Once the coefficients a and b are determined, the polarization internal resistances R_1 and R_2 can be calculated as follows:

$$R_1 = a/I, \tag{10}$$

$$R_2 = b/I, \tag{11}$$

where I is the measured current. The identification results for R_1 and R_2 are shown in Figure 6. It should be noted that the identification results for charging and discharging are different.

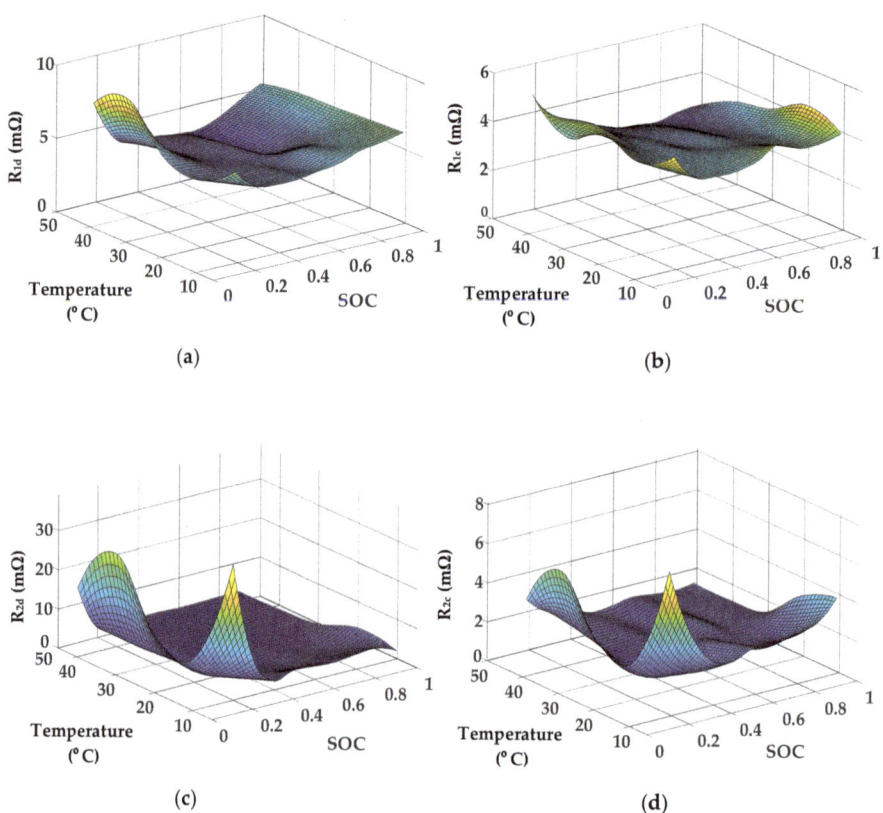

Figure 6. Polarization internal resistance identification results: (**a**) identified internal resistance for discharging, R_{1d}; (**b**) identified internal resistance for charging, R_{1c}; (**c**) identified internal resistance for discharging, R_{2d}; and (**d**) identified internal resistance for charging, R_{2c}.

After the internal resistances R_1 and R_2 are determined, the polarization capacitances C_1 and C_2 can be computed as follows:

$$C_1 = R_1/\tau_1, \tag{12}$$

$$C_2 = R_2/\tau_2. \tag{13}$$

The identification results for C_1 and C_2 are shown in Figure 7. Similar to the case of polarization internal resistances, the capacitance identification results for charging and discharging are different. It is clearly seen in Figures 6 and 7 that the values of these four parameters, R_1, R_2, C_1, and C_2, are dependent on SOC, ambient temperature, and direction of current (i.e., charging or discharging).

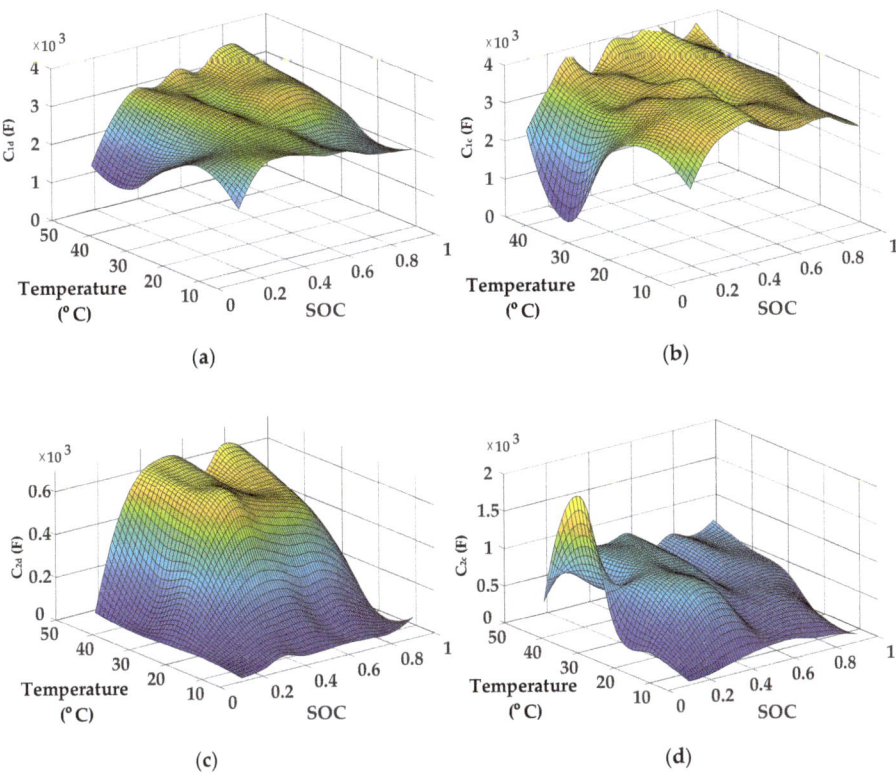

Figure 7. Polarization capacitance identification results: (**a**) identified capacitance for discharging, C_{1d}; (**b**) identified capacitance for charging, C_{1c}; (**c**) identified capacitance for discharging, C_{2d}; and (**d**) identified capacitance for charging, C_{2c}.

Substituting the voltage drop values (as shown in Figure 5) in Equations (14) and (15), the ohmic internal resistance R_0 in the battery model is obtained:

$$R_{0d} = ((U_1 - U_2) + (U_4 - U_3))/(2 \times I_d), \tag{14}$$

$$R_{0c} = ((U_6 - U_5) + (U_7 - U_8))/(2 \times I_c), \tag{15}$$

$$R_0 = (R_{0d} + R_{0c})/2. \tag{16}$$

where R_{0d} and R_{0c} represent the ohmic internal resistances during discharging and charging, respectively; and I_d and I_c denote the discharging current and charging current, respectively. The identification results for the ohmic internal resistance R_0, at different ambient temperatures, are shown in Table 4. It is seen that the value of R_0 drops as the ambient temperature increases.

Table 4. Identification results for the ohmic internal resistance, R_0.

Parameter	5 °C	15 °C	25 °C	35 °C	45 °C
R_0	0.067551	0.056435	0.048624	0.046138	0.043975

4. Model Verification and Discussion

In this section, the accuracy of the proposed temperature-dependent second-order RC equivalent circuit model is verified. The schematic of the simulation model used for verification is shown in Figure 8. The identified parameters obtained from Section 3 are employed in the proposed model, and the model accuracy is evaluated in terms of the difference (error) between the measured terminal voltage and that resulting from the proposed model.

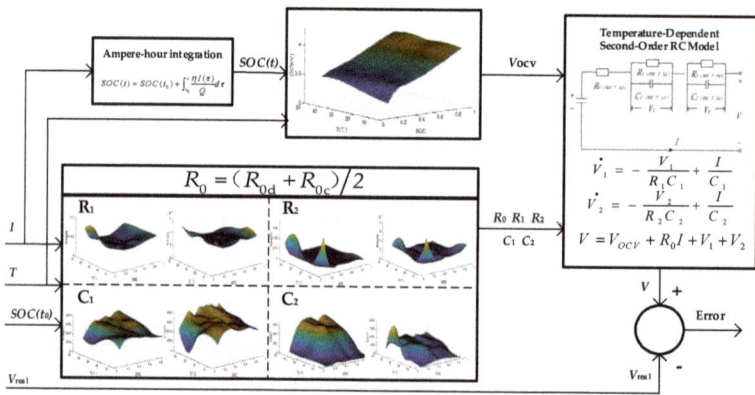

Figure 8. Schematic of the simulation model.

For verification purposes, the model has undergone the discharging test, charging test, and dynamic stress test (DST) [34], under constant and varying temperature conditions. The voltages resulting from the model as well as the voltage errors are plotted in Figure 9, under a constant ambient temperature of 25 °C. It is shown that for both the discharging and charging tests, the model output voltages are very close to the measured terminal voltages, and the voltage errors are maintained within ±20 mV. As for the DST, the model output voltage follows the measured voltage very well, with a slightly increased error magnitude compared with the first two cases. Note that the error is still maintained within a very small range, that is, ±50 mV.

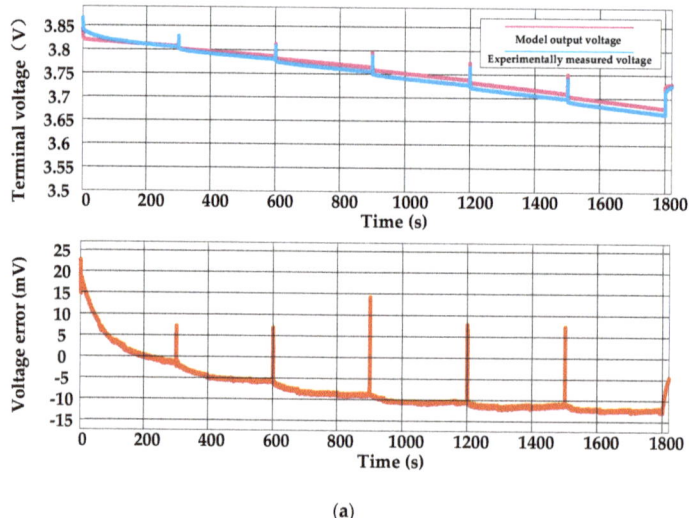

(a)

Figure 9. Cont.

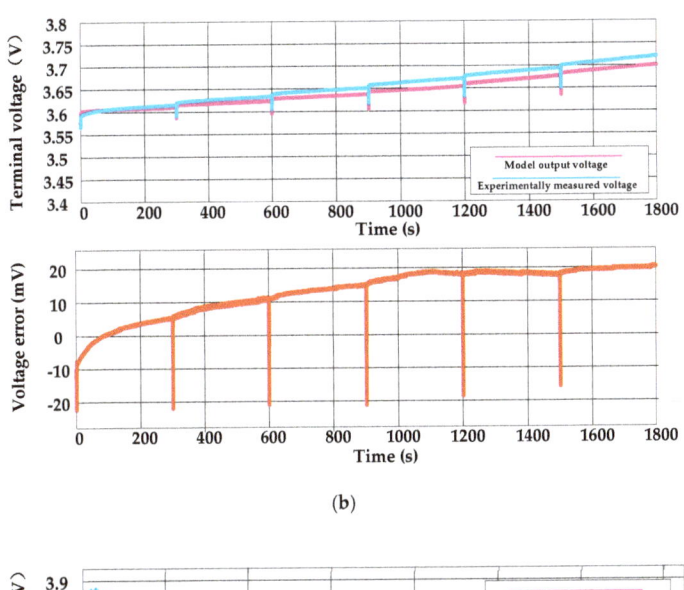

(b)

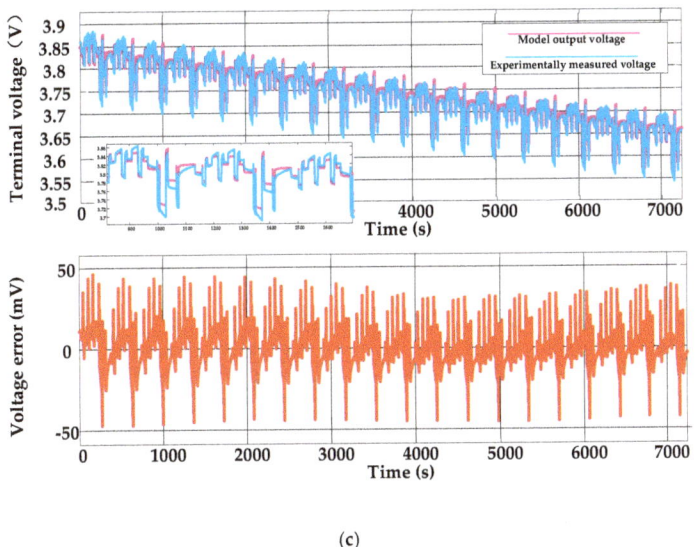

(c)

Figure 9. Model verification results at 25 °C: (**a**) model output voltage and its error for the discharging test, (**b**) model output voltage and its error for the charging test, and (**c**) model output voltage and its error for the dynamic stress test (DST). The magnified mini-plot is an enlarged view for the time range of 800 s–1800 s.

The verification results under varying temperature conditions are shown in Figure 10. In the discharging test, the ambient temperature was gradually increased from 12 °C to 20 °C. We see from Figure 10a that the model output voltage is quite close to the experimentally measured battery terminal voltage, with the voltage error remaining within ±15 mV. Besides, a traditional second-order RC model, which is independent of ambient temperature, is introduced for comparison purposes, and its output voltage is represented by an orange curve. It is clearly shown that the purple curve, generated by the proposed temperature-dependent second-order RC model, is closer to the measured voltage compared with the orange curve.

In the charging test, the ambient temperature gradually was decreased from 22 °C to 14 °C. As shown in Figure 10b, similar to the previous case, the model output voltage is close to the measured voltage and the voltage error is maintained within ±30 mV. Again, the purple curve, representing the proposed model, is closer to the measured voltage compared with the orange curve.

As for the DST, the ambient temperature was gradually increased from 25 °C to 40 °C. In this case, the model output voltage is still close to the measured battery terminal voltage, and the voltage error is kept within ±50 mV. Same as the above two cases, the proposed model outperforms traditional second-order RC model, as clearly demonstrated by the enlarged curves in Figure 10c.

The above results indicate that the established temperature-dependent second-order RC model provides not only accurate output voltage, but also good robustness against temperature variations. This verifies the superiority of the proposed model to the traditional second-order RC model under varying ambient temperature conditions.

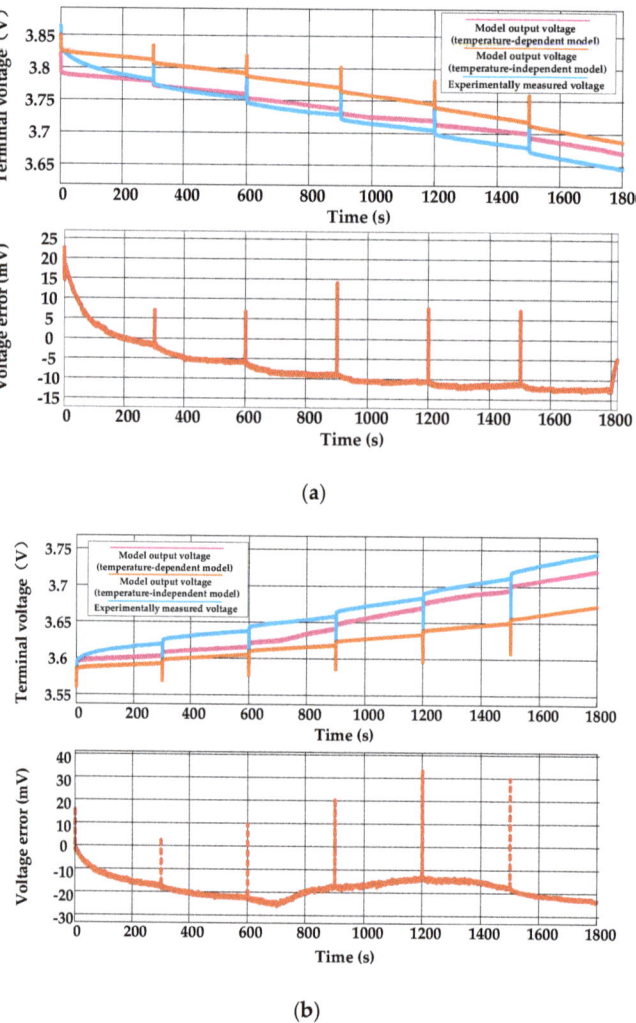

Figure 10. Cont.

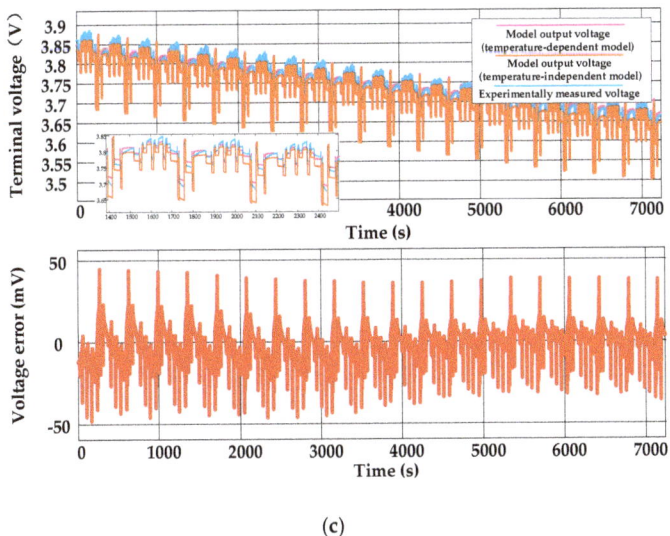

(c)

Figure 10. Model verification results under varying temperature conditions: (**a**) model output voltage and its error for the discharging test, (**b**) model output voltage and its error for the charging test, and (**c**) model output voltage and its error for the DST. The magnified mini-plot is an enlarged view for the time range of 1400 s–2400 s.

5. SOC Estimation Based on Double Kalman Filter Algorithm

It is known that high-precision SOC estimation is crucial for batteries' safety, power characteristics, and service life, and the accuracy of SOC estimation is directly dependent on the performance of the battery model used. In this section, a double Kalman filter (DKF) algorithm is proposed for SOC estimation, based on the established temperature-dependent second-order RC model. This algorithm combines the advantages of the Ah integration method, the KF algorithm, and the EKF algorithm, and its effectiveness is verified through comparisons with the EKF algorithm.

5.1. DKF Algorithm

The schematic that demonstrates the structure of the proposed DKF algorithm is shown in Figure 11. This algorithm employs a two-layer filtering layout to provide SOC estimation. In the first layer, the error between the measured terminal voltage (i.e., V_{real}) and that resulting from the proposed battery model (namely V) is employed as the input to an EKF. The output from the EKF is fed to the proposed battery model to produce a corrected SOC estimate, SOC_{EKF}. The purpose of this EKF is to deal with the uncertainties caused by modeling imperfections, thereby improving the SOC estimation performance. In the second layer, the error between SOC_{EKF} and that produced by the Ah integration method (denoted by SOC_{Ah}) is sent to a KF, and the output from this KF is then fed back to the Ah integration algorithm, which generates a further corrected SOC estimate, that is, the output SOC. The purpose of this KF is to suppress the cumulative error resulting from the Ah integration method, which further enhances the SOC estimation accuracy.

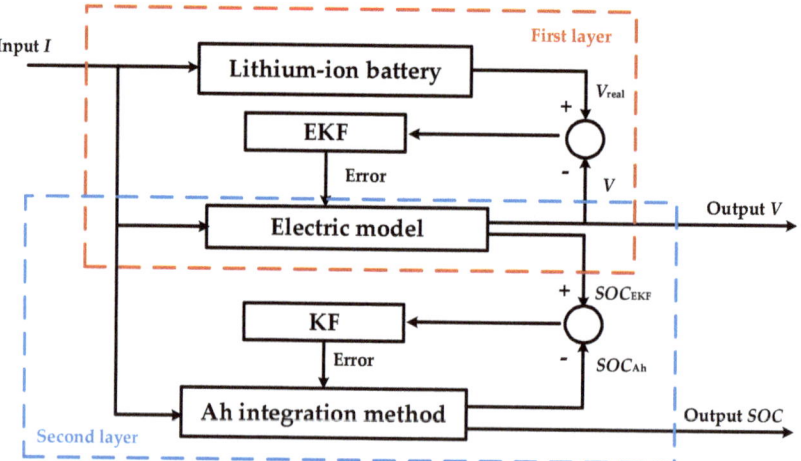

Figure 11. Schematic of the proposed dual Kalman filter (DKF) algorithm for SOC estimation. EKF, extended Kalman filter.

To implement the DKF algorithm, we first rewrite the proposed temperature-dependent second-order RC model in a discrete state space form. In this expression, the state vector is $x = [SOC\ V_1\ V_2]^T$, the input is $u = I$ (battery current), and the output is $y = V$ (battery terminal voltage). At time k, this state space equation takes the following form:

$$\begin{cases} x_{k+1} = Ax_k + Bu_k + \omega_k \\ y_k = Cx_k + Du_k + v_k \end{cases}, \tag{17}$$

where ω_k denotes the process noise; v_k represents the measurement noise; and matrices A, B, C, and D are given by the following:

$$A = \begin{bmatrix} 1 & 0 & 0 \\ 0 & \exp\left(\frac{-t}{R_1 C_1}\right) & 0 \\ 0 & 0 & \exp\left(\frac{-t}{R_1 C_1}\right) \end{bmatrix}, \tag{18}$$

$$B = \begin{bmatrix} \frac{\eta t}{Q} \\ R_1\left(1 - \exp\left(\frac{-t}{R_1 C_1}\right)\right) \\ R_2\left(1 - \exp\left(\frac{-t}{R_2 C_2}\right)\right) \end{bmatrix}, \tag{19}$$

$$C = \begin{bmatrix} \frac{dOCV}{dSOC} & 1 & 1 \end{bmatrix}, \tag{20}$$

$$D = R_0. \tag{21}$$

The detailed implementation process of the proposed DKF algorithm is as follows:

Step 1. Determination of the initial SOC value: Obtain the initial SOC value by means of the SOC–T–OCV map.

Step 2. State initialization: Determine the SOC error covariance, the process noise variance, and the measurement noise variance.

Step 3. First layer filtering: Conduct EKF-based filtering using Equations (18)–(21), and produce a corrected SOC estimate, SOC_{EKF}.

Step 4. Second layer filtering: Conduct KF-based filtering using Equations (18)–(21), and produce a further corrected SOC estimate, that is, the output SOC.

Step 5. Iteration: Repeat steps 3 and 4 to provide real-time SOC estimation.

Following the above recursive filtering steps, the uncertainties caused by modeling imperfections, as well as the cumulative error resulting from the Ah integration method, can both be suppressed. Through the proposed two-layer filtering mechanism, the overall SOC estimation performance is greatly improved, as we will show in the following section.

5.2. Algorithm Verification and Discussion

The effectiveness of the proposed DKF algorithm is verified by means of the DST. Similar to Section 4, the SOC estimation accuracy is evaluated under both constant and varying temperature conditions. Figure 12 shows the SOC estimation results produced by the proposed DKF algorithm and the traditional EKF algorithm. We see that for both temperature conditions, the black curve (representing the DKF) is closer to the real value, with significantly fewer pulses and ripples, compared with the blue curve produced by the EKF. Besides, it is also shown in this figure that the SOC estimation errors resulting from the proposed DKF are maintained within a very small range, ±0.4%, for both temperature conditions. The accuracy of the EKF and the proposed DKF under different working conditions is shown in Table 5.

Note that the proposed DKF algorithm is robust to the initial SOC error. In other words, the proposed DKF is able to ensure the convergence of the estimated SOC to the real value, even if an error exists in the initial SOC value. We see that in Figure 13a, the initial SOC value fed to the DKF is higher than the real value, while in Figure 13b, the initial SOC value is lower than the real one. In both cases, the proposed DKF algorithm gradually drives the estimated SOC towards the real values, and the estimation errors are kept within ±0.5% after convergence.

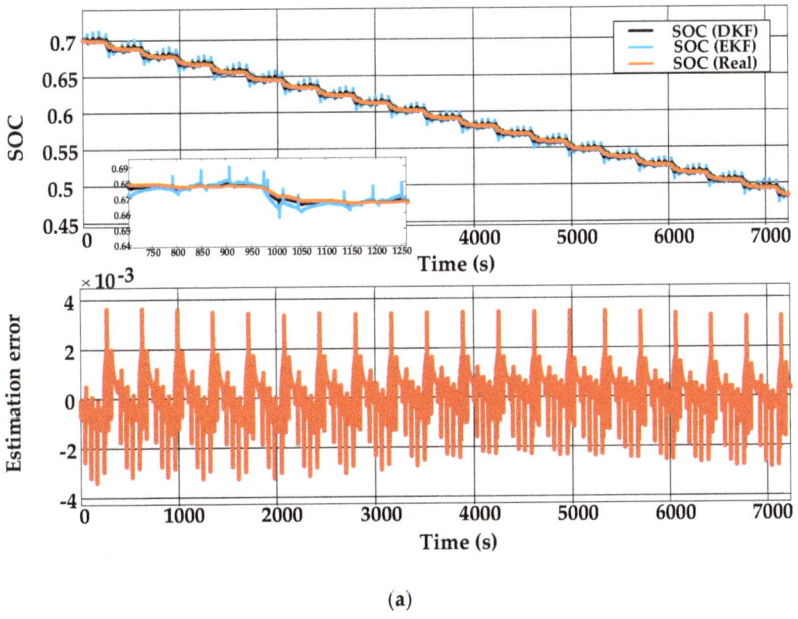

(a)

Figure 12. Cont.

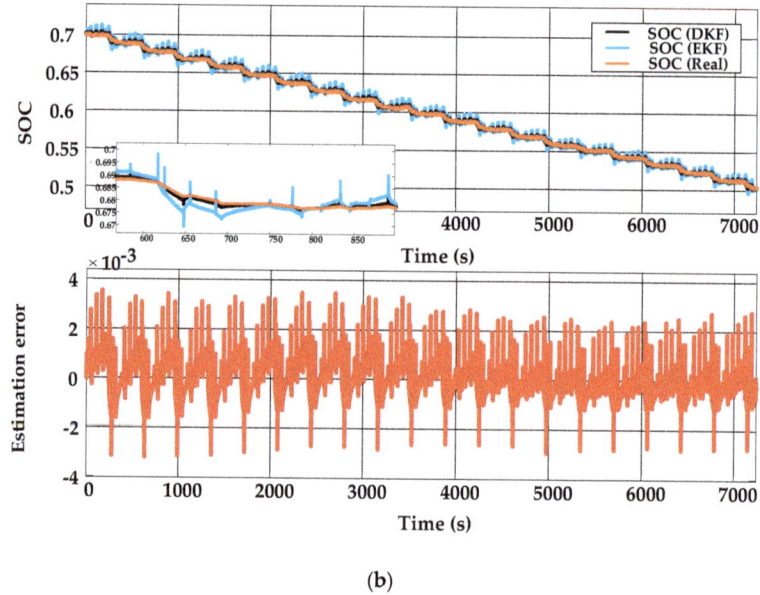

Figure 12. Algorithm verification results for the DST: (**a**) estimated SOC values and estimation error under constant temperature condition, the magnified mini-plot is an enlarged view for the time range of 550 s–900 s; and (**b**) estimated SOC values and estimation error under varying temperature condition. The magnified mini-plot is an enlarged view for the time range of 550 s–900 s.

Table 5. Accuracy comparison between extended Kalman filter (EKF) and dual Kalman filter (DKF) under different working conditions.

Working Conditions	EKF			DKF		
	R^2	Maximum error	Average error	R^2	Maximum error	Average error
Constant temperature	0.9978	0.0134	0.0024	0.9998	0.0036	0.0006
Varying temperature	0.9984	0.131	0.0018	0.9999	0.0035	0.0005

The above verification results indicate that the proposed DKF algorithm, constructed based on the established temperature-dependent second-order RC model, outperforms the traditional and classical EKF algorithm in terms of SOC estimation, under both constant and varying temperature conditions. In addition, the proposed DKF provides good robustness against initial SOC errors, which guarantees the performance of the DKF even if the initial SOC cannot be accurately obtained.

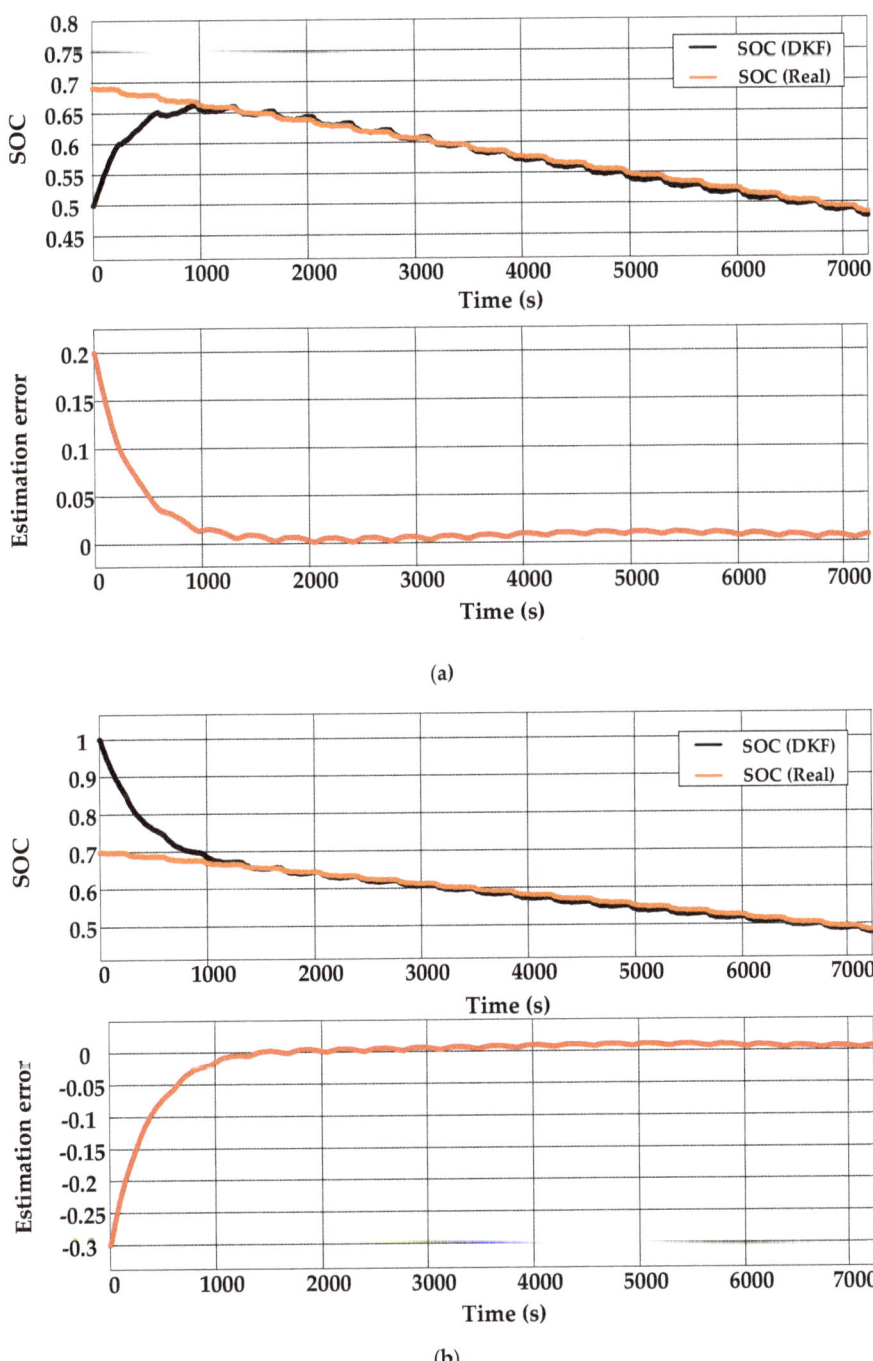

Figure 13. Algorithm verification results with initial SOC errors: (**a**) estimated SOC and corresponding estimation error with an initial SOC value of 0.5, (**b**) estimated SOC and corresponding estimation error with an initial SOC value of 1.

6. Conclusions

The battery electrical characteristics are dependent on the ambient temperature; however, most existing battery equivalent circuit models have not taken into account the influences of ambient temperature. In this paper, a temperature-dependent second-order RC equivalent circuit model is established, based on the electrical characteristics of lithium-ion batteries at different ambient temperatures (5 °C, 15 °C, 25 °C, 35 °C, and 45 °C). The unknown model parameters are identified by means of experiments, and the accuracy of the proposed battery model is verified under constant and varying temperature conditions. It is shown that for both the discharging and charging tests, the model errors are maintained within ±30 mV. As for the DST, the accuracy of the model is reduced, but the error is still maintained within a very small range, that is, ±50 mV. These verification results indicate that the proposed model provides not only accurate output voltage, but also good robustness against temperature variations. Besides, it is also shown that the proposed model outperforms the traditional second-order RC model under varying ambient temperature conditions.

On the basis of the proposed temperature-dependent second-order RC model, a DKF algorithm that combines the Ah integration method, the KF algorithm, and the EKF algorithm is proposed for SOC estimation. This algorithm employs a two-layer filtering mechanism to enhance SOC estimation performance, which suppresses the uncertainties caused by modeling imperfections, as well as the cumulative error resulting from the Ah integration method. The effectiveness of the proposed DKF algorithm is verified by means of the DST, under both constant and varying temperature conditions. The verification results indicate that the proposed DKF algorithm outperforms the traditional and classical EKF algorithm, with the SOC estimation errors under different temperature conditions maintained within ±0.4%. Besides, it is also proven that the proposed DKF algorithm provides good robustness against temperature variations and initial SOC errors.

Lithium-ion batteries will age with the increase of the number of cycles. Consequently, the parameters of the battery model will also change with time. In our future work, the battery lifetime experiments will be performed to obtain battery aging data, based on which a more advanced battery model considering battery aging will be investigated and utilized for better battery state estimation.

Author Contributions: Conceptualization, M.H.; methodology, M.H.; validation, Z.S.; formal analysis, Y.X.; investigation, Y.X., K.C., Z.S. and Z.Y.; writing—original draft preparation, Y.X.; writing—review and editing, C.F.; supervision, M.H. and C.F.; project administration, K.C. and Z.Y.; funding acquisition, M.H. and C.F.

Funding: This research was funded by the National Key Research and Development Project, grant number 2018YFB0106102.

Conflicts of Interest: The authors declare no conflict of interest. The funders had no role in the design of the study; in the collection, analyses, or interpretation of data; in the writing of the manuscript; or in the decision to publish the results.

References

1. Chaoui, H.; Ibe-Ekeocha, C.C. State of Charge and State of Health Estimation for Lithium Batteries Using Recurrent Neural Networks. *IEEE Trans. Veh. Technol.* **2017**, *66*, 8773–8783. [CrossRef]
2. Ng, K.S.; Moo, C.S.; Chen, Y.P.; Hsieh, Y.C. Enhanced coulomb counting method for estimating state-of-charge and state-of-health of lithium-ion batteries. *Appl. Energy* **2009**, *86*, 1506–1511. [CrossRef]
3. Stan, A.I.; Swierczynski, M.; Stroe, D.I.; Teodorescu, R.; Andreasen, S.J. Lithium Ion Battery Chemistries from Renewable Energy Storage to Automotive and Back-up Power Applications—An Overview. In Proceedings of the 2014 International Conference on Optimization of Electrical and Electronic Equipment (Optim), Bran, Romania, 22–24 May 2014; pp. 713–720.
4. Xiong, R.; Cao, J.Y.; Yu, Q.Q.; He, H.W.; Sun, F.C. Critical Review on the Battery State of Charge Estimation Methods for Electric Vehicles. *IEEE Access* **2018**, *6*, 1832–1843. [CrossRef]
5. Yang, N.X.; Zhang, X.W.; Shang, B.B.; Li, G.J. Unbalanced discharging and aging due to temperature differences among the cells in a lithium-ion battery pack with parallel combination. *J. Power Sources* **2016**, *306*, 733–741. [CrossRef]

6. Deng, Y.; Hu, Y.L.; Cao, Y. An Improved Algorithm of SOC Testing Based on Open-Circuit Voltage-Ampere Hour Method. *Intell. Comput. Smart Grid Electr. Veh.* **2014**, *463*, 258–267.
7. Lee, S.; Kim, J.; Lee, J.; Cho, B.H. State-of-charge and capacity estimation of lithium-ion battery using a new open-circuit voltage versus state-of-charge. *J. Power Sources* **2008**, *185*, 1367–1373. [CrossRef]
8. Orchard, M.E.; Hevia-Koch, P.; Zhang, B.; Tang, L. Risk Measures for Particle-Filtering-Based State-of-Charge Prognosis in Lithium-Ion Batteries. *IEEE Trans. Ind. Electron.* **2013**, *60*, 5260–5269. [CrossRef]
9. Piller, S.; Perrin, M.; Jossen, A. Methods for state-of-charge determination and their applications. *J. Power Sources* **2001**, *96*, 113–120. [CrossRef]
10. Tulsyan, A.; Tsai, Y.; Gopaluni, R.B.; Braatz, R.D. State-of-charge estimation in lithium-ion batteries: A particle filter approach. *J. Power Sources* **2016**, *331*, 208–223. [CrossRef]
11. Plett, G.L. Extended Kalman filtering for battery management systems of LiPB-based HEV battery packs. Part 3. State and parameter estimation. *J. Power Sources* **2004**, *134*, 277–292. [CrossRef]
12. Plett, G.L. Sigma-point Kalman filtering for battery management systems of LiPB-based HEV battery packs—Part 2: Simultaneous state and parameter estimation. *J. Power Sources* **2006**, *161*, 1369–1384. [CrossRef]
13. Sun, F.C.; Xiong, R.; He, H.W. A systematic state-of-charge estimation framework for multi-cell battery pack in electric vehicles using bias correction technique. *Appl. Energy* **2016**, *162*, 1399–1409. [CrossRef]
14. Waag, W.; Fleischer, C.; Sauer, D.U. Critical review of the methods for monitoring of lithium-ion batteries in electric and hybrid vehicles. *J. Power Sources* **2014**, *258*, 321–339. [CrossRef]
15. Zheng, Y.J.; Ouyang, M.G.; Han, X.B.; Lu, L.G.; Li, J.Q. Investigating the error sources of the online state of charge estimation methods for lithium-ion batteries in electric vehicles. *J. Power Sources* **2018**, *377*, 161–188. [CrossRef]
16. Cai, C.H.; Du, D.; Liu, Z.Y.; Zhang, H. Artificial neural network in estimation of battery state-of-charge (SOC) with nonconventional input variables selected by correlation analysis. In Proceedings of the 2002 International Conference on Machine Learning and Cybernetics, Beijing, China, 4–5 November 2002; Volume 3, pp. 1619–1625.
17. He, H.W.; Zhang, X.W.; Xiong, R.; Xu, Y.L.; Guo, H.Q. Online model-based estimation of state-of-charge and open-circuit voltage of lithium-ion batteries in electric vehicles. *Energy* **2012**, *39*, 310–318. [CrossRef]
18. Shi, Q.-S.; Zhang, C.-H.; Cui, N.-X. Estimation of battery state-of-charge using ν-support vector regression algorithm. *Int. J. Automot. Technol.* **2008**, *9*, 759–764. [CrossRef]
19. Schwunk, S.; Armbruster, N.; Straub, S.; Kehl, J.; Vetter, M. Particle filter for state of charge and state of health estimation for lithium–iron phosphate batteries. *J. Power Sources* **2013**, *239*, 705–710. [CrossRef]
20. Khumprom, P.; Yodo, N. A data-driven predictive prognostic model for lithium-ion batteries based on a deep learning algorithm. *Energies* **2019**, *12*, 660. [CrossRef]
21. Li, C.; Xiao, F.; Fan, Y. An approach to state of charge estimation of lithium-ion batteries based on recurrent neural networks with gated recurrent unit. *Energies* **2019**, *12*, 1592. [CrossRef]
22. Zenati, A.; Desprez, P.; Razik, H. Estimation of the SOC and the SOH of Li-ion Batteries, by combining Impedance Measurements with the Fuzzy Logic Inference. In Proceedings of the 36th Annual Conference on IEEE Industrial Electronics Society (IECON 2010), Glendale, AZ, USA, 7–10 November 2010; pp. 1773–1778.
23. Bhangu, B.S.; Bentley, P.; Stone, D.A.; Bingham, C.M. Nonlinear Observers for Predicting State-of-Charge and State-of-Health of Lead-Acid Batteries for Hybrid-Electric Vehicles. *IEEE Trans. Veh. Technol.* **2005**, *54*, 783–794. [CrossRef]
24. Chen, Z.; Fu, Y.H.; Mi, C.C. State of Charge Estimation of Lithium-Ion Batteries in Electric Drive Vehicles Using Extended Kalman Filtering. *IEEE Trans. Veh. Technol.* **2013**, *62*, 1020–1030. [CrossRef]
25. Li, S.X.; Hu, M.H.; Li, Y.X.; Gong, C.C. Fractional-order modeling and SOC estimation of lithium-ion battery considering capacity loss. *Int. J. Energy Res.* **2019**, *43*, 417–429. [CrossRef]
26. Duong, V.H.; Bastawrous, H.A.; Lim, K.; See, K.W.; Zhang, P.; Dou, S.X. Online state of charge and model parameters estimation of the LiFePO4 battery in electric vehicles using multiple adaptive forgetting factors recursive least-squares. *J. Power Sources* **2015**, *296*, 215–224. [CrossRef]
27. Yang, F.F.; Xing, Y.J.; Wang, D.; Tsui, K.L. A comparative study of three model-based algorithms for estimating state-of-charge of lithium-ion batteries under a new combined dynamic loading profile. *Appl. Energy* **2016**, *164*, 387–399. [CrossRef]

28. Rooij, D.M.R.D. Electrochemical Methods: Fundamentals and Applications. *Anti Corros. Methods Mater.* **2003**, *50*, 414–421. [CrossRef]
29. Yatsui, M.W.; Bai, H. Kalman filter based state-of-charge estimation for lithium-ion batteries in hybrid electric vehicles using pulse charging. In Proceedings of the 2011 IEEE Vehicle Power and Propulsion Conference, Chicago, IL, USA, 6–9 September 2011.
30. Hu, M.H.; Li, Y.X.; Li, S.X.; Fu, C.Y.; Qin, D.T.; Li, Z.H. Lithium-ion battery modeling and parameter identification based on fractional theory. *Energy* **2018**, *165*, 153–163. [CrossRef]
31. Idaho National Engineering and Environmental Laboratory. *FreedomCAR Battery Test Manual for Power-Assist Hybrid Electric Vehicles*; INEEL (DOE/ID-11069): Cuernavaca, Mexico, 2003.
32. Zhu, Q.; Xiong, N.; Yang, M.L.; Huang, R.S.; Hu, G.D. State of Charge Estimation for Lithium-Ion Battery Based on Nonlinear Observer: An H-infinity Method. *Energies* **2017**, *10*, 679. [CrossRef]
33. Zhu, Q.; Li, L.; Hu, X.S.; Xiong, N.; Hu, G.D. H-infinity-Based Nonlinear Observer Design for State of Charge Estimation of Lithium-Ion Battery with Polynomial Parameters. *IEEE Trans. Veh. Technol.* **2017**, *66*, 10853–10865. [CrossRef]
34. Idaho National Engineering Laboratory. *USABC Electric Vehicle Battery Test Procedures Manual–Revision 2*; INEEL (DOE/ID-10479): Cuernavaca, Mexico, 1996.

© 2019 by the authors. Licensee MDPI, Basel, Switzerland. This article is an open access article distributed under the terms and conditions of the Creative Commons Attribution (CC BY) license (http://creativecommons.org/licenses/by/4.0/).

Article

A Method to Identify Lithium Battery Parameters and Estimate SOC Based on Different Temperatures and Driving Conditions

Yongliang Zheng, Feng He * and Wenliang Wang

Department of Automotive Engineering, School of Mechanical Engineering, Guizhou University, Guiyang 550025, China; gs.ylzheng17@gzu.edu.cn (Y.Z.); gs.wangwl18@gzu.edu.cn (W.W.)
* Correspondence: hef@gzu.edu.cn

Received: 2 November 2019; Accepted: 18 November 2019; Published: 22 November 2019

Abstract: State of charge (SOC) plays a significant role in the battery management system (BMS), since it can contribute to the establishment of energy management for electric vehicles. Unfortunately, SOC cannot be measured directly. Various single Kalman filters, however, are capable of estimating SOC. Under different working conditions, the SOC estimation error will increase because the battery parameters cannot be estimated in real time. In order to obtain a more accurate and applicable SOC estimation than that of a single Kalman filter under different driving conditions and temperatures, a second-order resistor capacitor (RC) equivalent circuit model (ECM) of a battery was established in this paper. Thereafter, a dual filter, i.e., an unscented Kalman filter–extended Kalman filter (UKF–EKF) was developed. With the EKF updating battery parameters and the UKF estimating the SOC, UKF–EKF has the ability to identify parameters and predict the SOC of the battery simultaneously. The dual filter was verified under two different driving conditions and three different temperatures, and the results showed that the dual filter has an improvement on SOC estimation.

Keywords: state of charge; battery parameters identification; equivalent circuit model; dual Kalman filter

1. Introduction

With the increasing energy crisis, alternative energy vehicles have been given full attention. Lithium-ion batteries (LIBs) have become the power source of electric vehicles (EVs) because of their high energy density and long service life [1]. Battery management systems (BMS) can predict the driving mileage of EVs through the state of charge (SOC) of LIBs, and they can formulate the energy management strategies of EVs, which can not only extend the life of batteries, but the driving mileage of cars, as well. The monitoring of SOC also plays a significant role in preventing batteries in EVs from some dangerous operations, like overcharging or overdischarging.

Unfortunately, unlike other parameters of the battery, such as current and terminal voltage, SOC cannot be measured directly. Many scholars have studied SOC estimation for a long time; below are some common methods of SOC estimation:

(i) Although the Coulomb counting method (CCM) is uncomplicated and has been applied in SOC prediction, in practice, error accumulation will gradually increase due to the influence of uncertain noise. In addition, the initial value of the SOC will also affect the estimation accuracy, so it is not appropriate for the SOC estimation of LIBs in EVs.

(ii) Through the open-circuit voltage method (OCVM), a relatively functional relationship between the open-circuit voltage (OCV) and the SOC can be found; hence, SOC can be estimated according to the changing OCV at work. It is, however, limited to estimating the SOC in LIBs for the need of long rest and the neglect of changes in the internal resistance of LIBs during working.

(iii) The improvements in computer computing ability are helpful in the popularization and application of machine learning (ML). Some ML algorithms, such as neural networks (NN) [2,3], extreme learning machines (EXL) [4,5], and support vector machines (SVM) [6], only need to train some input parameters, such as terminal voltage, operating current, resistance, and the temperatures of LIBs during working, to predict SOC. However, because of the constraints of sample training and the large amounts of computation required, it is difficult to meet the requirements of BMS to predict SOC at this time.

(iv) Based on an electrical equivalent circuit model (ECCM), some advanced Kalman filter (KF) methods, like the extended Kalman filter (EKF) [7,8], the unscented Kalman filter (UKF), [9] and the cubature Kalman filter (CKF), [10] have demonstrated their abilities to accurately predict SOC. Moreover, in order to deal with the uncertainties of system noise in SOC estimation, some Kalman filters improved by adaptive algorithms, such as the adaptive extended Kalman filter (AEKF) [11], the adaptive unscented Kalman filter (AUKF), [12] and the adaptive cubature Kalman filter (ACKF) [13], have been used to improve the estimation accuracy of SOC. Nevertheless, the resistances of lithium batteries vary in different working conditions and temperatures, which are neglected when the SOC is estimated by a single Kalman filter. As a consequence, the estimation error of SOC will increase. It is a common approach to identify battery parameters by the hybrid pulse power characteristic (HPPC) offline test, but the accuracy of this method is not guaranteed, and it does consume a lot of time.

In order to resolve the aforementioned shortcomings of the single Kalman filter, many scholars have proposed joint estimation methods to simultaneously achieve SOC estimation and battery parameter identification online. Xiong et al. [14] proposed a joint estimator to predict SOC and state of power (SOP) capabilities, with recursive least squares (RLS) successfully updating battery parameters and the AEKF achieving the estimation of the SOC and power capacity. Wei et al. [15] compared three methods—the dual extended Kalman filter (DEKF), the recursive least squares–extended Kalman filter (RLS–EKF), and the noise compensating EKF (NC-EKF)—for battery parameter identification and SOC online estimation. All of them had high accuracy for SOC estimation and parameter identification under low noise interference, but as the unknown interference increased, they showed their own weaknesses in calculating cost, number of parameters to be adjusted, and robustness of estimation. These SOC estimators are only based on the first-order RC ECM, and RLS is not suitable for estimating nonlinear systems. EKF also suffers from low SOC estimation accuracy for abandoning higher order. The establishment of UKF–EKF by Zhang et al. [1] has a good ability to predict the SOC and the parameters of the battery pack, but its stability has not been verified at different temperatures.

In fact, in addition to the unknown noise at work, the estimation of the SOC and identification of the parameters of the battery are usually influenced by ambient temperature, which is reflected in the changes of the OCV–SOC function relationship and the battery's internal resistance. In order to realize SOC estimation and online parameter identification of the battery simultaneously, this paper proposes a second-order RC ECM of the battery, and a joint estimation model, i.e., the UKF–EKF is structured. The UKF is used to predict the SOC, which solves the shortcomings of insufficient accuracy of the EKF's SOC estimation, while the EKF updates the battery parameters, which further enhances the accuracy of the SOC estimation.

2. Battery of the Equivalent Circuit Model

The successful application of the Kalman filter in SOC estimation depends on the accurate model of the battery. Many studies [16–18] have shown that the ECM can be successfully used in SOC estimation because it can accurately reflect the physical and chemical changes of the battery, and the computing cost is low, which conforms to the requirements of a BMS. Considering the accuracy and computational complexity of the battery model, a second-order ECM is selected and plotted in Figure 1.

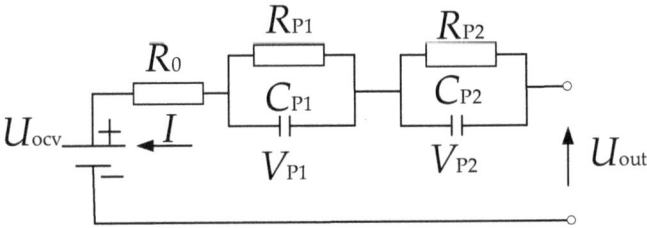

Figure 1. Second-order resistor capacitor (RC) equivalent circuit model (ECM) of a battery.

I is the working current, which is positive when discharging and negative when charging. R_0 represents ohmic resistance. R_{P1} and C_{P1} are resistance and capacitance of electrochemical polarization, respectively, while R_{P2} and C_{P2} represent the resistance and capacitance of concentration polarization, respectively. U_{ocv} and U_{out} are open-circuit voltage and terminal voltage, respectively. V_{P1} and V_{P2} are polarization voltage, and the state equation of the battery can be expressed by Equation (1):

$$\begin{cases} \dot{V}_{P1} = -\frac{1}{R_{P1}C_{P1}}V_{P1} + \frac{1}{C_{P1}}I \\ \dot{V}_{P2} = -\frac{1}{R_{P2}C_{P2}}V_{P2} + \frac{1}{C_{P2}}I \\ U_{out} = U_{ocv} - V_{P1} - V_{P2} - IR_0 \end{cases} \quad (1)$$

3. Dual Kalman Filter Design

In order to overcome the disadvantages of the traditional single unscented Kalman filter, which regards internal resistance as a constant, it is an advisable method to apply the dual Kalman filter in the joint online estimation of battery parameters and SOC. The UKF–EKF consists of two running parallel KFs, i.e., EKF is employed in the identification of battery parameters, while UKF is applied to SOC estimation.

The SOC can be expressed by the following formula:

$$SOC = SOC_0 + \frac{\Delta t i(t)}{C_p}, \quad (2)$$

where SOC_0 is the initial SOC, and C_p and $i(t)$ represent the rated capacity of the batteries and current, respectively.

Combine Equations (1) and (2), and a new equation of state with battery parameters can be obtained.

$$\begin{cases} X_{k+1} = \begin{bmatrix} 1 & 0 & 0 \\ 0 & e^{-\frac{\Delta t}{R_{P1}C_{P1}}} & 0 \\ 0 & 0 & e^{-\frac{\Delta t}{R_{P2}C_{P2}}} \end{bmatrix} \times X_k + \begin{bmatrix} -\Delta t/C_p \\ R_{P1}(1 - e^{\frac{\Delta t}{R_{P1}C_{P1}}}) \\ R_{P2}(1 - e^{\frac{\Delta t}{R_{P2}C_{P2}}}) \end{bmatrix} \times I_k + w_k^x \\ \theta_{k+1} = \theta_k + w_k^\theta \end{cases} \quad (3)$$

$$U_{out,k} = U_{ocv}(SOC_k) - V_{P1,k} - V_{P2,k} - I_k R_0 + v_k, \quad (4)$$

where the state vector $X_k = [SOC_k\ V_{P1,k}\ V_{P2,k}]^T$, $\tau_1 = R_{P1}C_{P1}$, $\tau_2 = R_{P2}C_{P2}$, and τ_{P1} and τ_{P2} are time constants in ECM. OCV can be obtained by fitting the relationship with SOC. $w_k^x \sim N(0, Q^x)$ and $w_k^\theta \sim N(0, Q^\theta)$ are process noise for state and parameters, respectively, measurement noise is $v_k \sim N(0, R^x)$, and battery parameters are $\theta = [R_0, R_{P1}, C_{P1}, R_{P2}, C_{P2}]^T$.

To obtain a nonlinear system with state and parameters, Equations (3) and (4) can be rewritten as:

$$x_{k+1} = f(x_k, \theta_k, u_k) + w_k^x = \theta_k + w_k^\theta \quad (5)$$

$$y_k = g(x_k, \theta_k, u_k) + v_k, \tag{6}$$

where the control variable $u_k = I_k$.

The detailed steps of UKF–EKF are shown below, and the flow chart is shown in Figure 2.

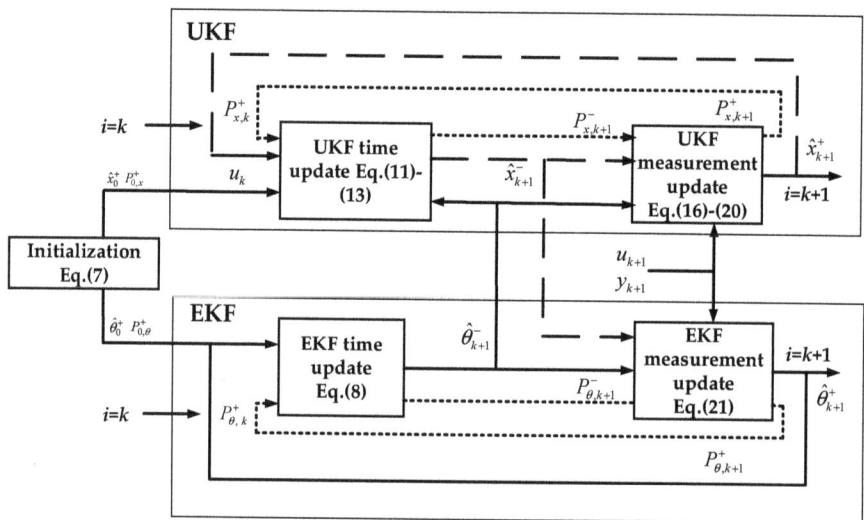

Figure 2. Flow chart of the unscented Kalman filter–extended Kalman filter (UKF–EKF).

(1) Initialization:

$$\begin{aligned}
\hat{x}_0^+ &= E[x_0] \\
P_{0,x}^+ &= E[(x_0 - \hat{x}_0^+)(x_0 - \hat{x}_0^+)]^T \\
\hat{\theta}_0^+ &= E[\hat{\theta}_0] \\
P_{0,\theta}^+ &= E[(\theta_0 - \hat{\theta}_0^+)(\theta_0 - \hat{\theta}_0^+)]^T
\end{aligned} \tag{7}$$

(2) Time update for battery parameters in EKF:

$$\begin{aligned}
\hat{\theta}_{k+1}^- &= \hat{\theta}_k^+ \\
P_{\theta,k+1}^- &= P_{\theta,k}^+ + Q_{\theta,k}
\end{aligned} \tag{8}$$

(3) Sigma sampling point and weight calculate for UKF:

$$\begin{cases}
x_{k|k}^0 = \hat{x}_{k|k}, i = 0 \\
x_{k|k}^i = \hat{x}_{k|k} + \left(\sqrt{(n+\lambda)P_{x,k}^+} \right)_i, i = 1 \sim n \\
x_{k|k}^i = \hat{x}_{k|k} - \left(\sqrt{(n+\lambda)P_{x,k}^+} \right)_i, i = n+1 \sim 2n
\end{cases} \tag{9}$$

$$\begin{cases}
\omega_m^{(0)} = \frac{\lambda}{n+\lambda} \\
\omega_c^{(0)} = \frac{\lambda}{n+\lambda} + (1 - \alpha^2 + \beta) \\
\omega_m^{(i)} = \omega_c^{(i)} = \frac{\lambda}{2(n+\lambda)}, i = 1 \sim 2n
\end{cases}, \tag{10}$$

where n is the state dimension of the battery. $\lambda = \alpha^2(n+k)-n$, the function of α is to control the distribution of sampling points, and the parameter k usually guarantees that the variance matrix is semipositive definite. β is the non-negative weight coefficient; in this study, $n = 3$, $\alpha = 0.01$, $k = 0$, $\beta = 2$.

(4) State estimation and error covariance time update:

$$x_{k+1|k}^i = f(x_{k|k}^i) \tag{11}$$

$$\hat{x}_{k+1|k} = \sum_{i=0}^{2n} \omega^{(i)} x_{k+1|k}^i \tag{12}$$

$$P_{k+1|k} = \sum_{i=0}^{2n} \omega^{(i)} [\hat{x}_{k+1|k} - x_{k+1|k}^i][\hat{x}_{k+1|k} - x_{k+1|k}^i]^T + Q_{x,k}. \tag{13}$$

(5) Update measurement with posteriori estimation:

$$Z_{k+1|k}^i = g(x_{k+1|k}^i) \tag{14}$$

$$\hat{Z}_{k+1|k}^i = \sum_{i=0}^{2n} \omega^{(i)} \hat{Z}_{k+1|k}^i. \tag{15}$$

(6) Update measurement covariance:

$$P_{y,k} = \sum_{i=0}^{2n} \omega^{(i)} [Z_{k+1|k}^i - \hat{Z}_{k+1|k}^i][Z_{k+1|k}^i - \hat{Z}_{k+1|k}^i]^T + R_{x,k} \tag{16}$$

$$P_{x,k} = \sum_{i=0}^{2n} \omega^{(i)} [x_{k+1|k}^i - \hat{Z}_{k+1|k}^i][Z_{k+1|k}^i - \hat{Z}_{k+1|k}^i]^T. \tag{17}$$

(7) Calculate UKF gain matrix:

$$K_{k+1} = P_{x,k}(P_{y,k})^{-1}. \tag{18}$$

(8) State estimation and error covariance measurement update:

$$\hat{x}_{k+1|k+1} = \hat{x}_{k+1|k} + K_{k+1}[Z_{k+1}^i - \hat{Z}_{k+1|k}^i] \tag{19}$$

$$P_{k+1|k+1} = P_{k+1|k} - K_{k+1} P_{z,k} (K_{k+1})^T. \tag{20}$$

(9) EKF measurement update for battery parameters:

$$\begin{aligned} K_{k,\theta} &= P_{\theta,k+1}^{-} (\hat{H}_{k,\theta})^T (\hat{H}_{k,\theta} P_{\theta,k+1}^{-} (\hat{H}_{k,\theta})^T + R_{k,\theta})^{-1} \\ \hat{\theta}_{k+1}^{+} &= \hat{\theta}_{k+1}^{-} + K_{k,\theta}(y_k - \hat{y}_k) \\ P_{\theta,k+1}^{+} &= (I - K_{k,\theta} \hat{H}_{k,\theta}) P_{\theta,k+1}^{-} \end{aligned} \tag{21}$$

The Jacobian equation of battery parameters is as follows:

$$\hat{H}_k^\theta = \left. \frac{dg(\hat{x}_{k+1}^-, u_{k+1}, 0)}{d\theta} \right|_{\theta = \hat{\theta}_{k+1}^-} = \frac{\partial g(\hat{x}_{k+1}^-, u_{k+1}, \theta)}{\partial \theta} + \frac{\partial g(\hat{x}_{k+1}^-, u_{k+1}, \theta)}{\partial \hat{x}_{k+1}^-} \frac{d\hat{x}_{k+1}^-}{d\theta} \tag{22}$$

$$\frac{d\hat{x}_{k+1}^-}{d\theta} = \frac{\partial f(\hat{x}_k^+, u_k, \theta)}{\partial \theta} + \frac{\partial f(\hat{x}_k^+, u_k, \theta)}{\partial \hat{x}_k^+} \frac{d\hat{x}_k^+}{d\theta} \tag{23}$$

$$\frac{d\hat{x}_k^+}{d\theta} = \frac{d\hat{x}_k^-}{d\theta} - K_k^x \frac{dg(\hat{x}_k^-, u_k, \theta)}{d\theta}, \tag{24}$$

where

$$\frac{\partial g(\hat{x}^-_{k+1}, u_{k+1}, \theta)}{\partial \theta} = [-I_{k+1}, 0, 0, 0, 0] \quad (25)$$

$$\frac{\partial g(\hat{x}^-_{k+1}, u_{k+1}, \theta)}{\partial \hat{x}^-_{k+1}} = [\partial U_{ocv}/\partial SOC_{k+1}, -1, -1] \quad (26)$$

$$\frac{\partial f(\hat{x}^+_k, u_k, \theta)}{\partial \theta} = \begin{bmatrix} 0 & 0 & 0 & 0 & 0 \\ 0 & a_{2,2} & a_{2,3} & 0 & 0 \\ 0 & 0 & 0 & a_{3,4} & a_{3,5} \end{bmatrix} \quad (27)$$

$a_{2,2} = V^-_{P1,k} \frac{\Delta t}{R^2_{P1} C_{P1}} \exp(-\Delta t/\tau_{P1}) - I_k(\exp(-\Delta t/\tau_{P1}) - 1) - \frac{I_k \Delta t}{R_{P1} C_{P1}} \exp(-\Delta t/\tau_{P1}),$

$a_{2,3} = (\Delta t/R_{P1} C^2_{P1})(V^-_{P1,k} - R_{P1} I_k) \exp(-\Delta t/\tau_{P1}),$

$a_{3,4} = V^-_{P2,k} \frac{\Delta t}{R^2_{P2} C_{P2}} \exp(-\Delta t/\tau_{P2}) - I_k(\exp(-\Delta t/\tau_{P2}) - 1) - \frac{I_k \Delta t}{R_{P2} C_{P2}} \exp(-\Delta t/\tau_{P2}),$

$a_{3,5} = (\Delta t/R_{P2} C^2_{P2})(V^-_{P2,k} - R_{P2} I_k) \exp(-\Delta t/\tau_{P2}).$

4. Experimental Design

The dataset of the 18650 battery was collected from an experiment conducted by Zheng et al. [19]. Battery data profiles consist of incremental current OCV and dynamic test profiles. The first profile data can be used to obtain the relationship between OCV and SOC, and Figure 3 plots OCV–SOC in incremental current OCV at 0, 25, and 45 °C, corresponding to low temperature, room temperature, and high temperature, respectively. It can be seen that at different temperatures, the OCV–SOC curves are different, which indicates that the electrode characteristics of the battery are influenced by temperature, which will affect the SOC estimation. The function of dynamic test profiles is to verify the ability of the UKF–EKF to achieve SOC estimation and parameter identification under different temperatures. There are two battery test loading profiles, including the highway condition, Highway Driving Schedule (US06), and the urban condition, Beijing Dynamic Stress Test (BJDST) in Figure 4. As can be seen from Figure 4, the charging/discharging current under the US06 working condition is larger; in addition, this working condition is more complex than the BJDST. The prediction results under different working conditions are shown in Section 5; the detailed parameters of the battery are shown in Table 1.

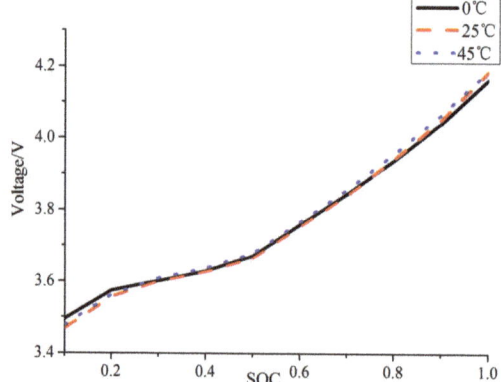

Figure 3. The relation diagram of open-circuit voltage (OCV)–state of charge (SOC) at different temperatures.

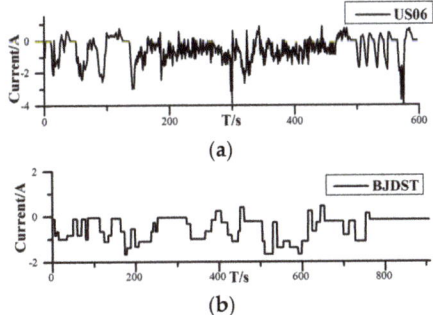

Figure 4. Profiles of battery test loading: (**a**) Highway Driving Schedule (US06); (**b**) Beijing Dynamic Stress Test (BJDST).

Table 1. Battery parameters.

Type	18650
Normal Voltage	3.6 V
Normal Capacity	2 Ah
Upper/lower cut-off voltage	4.2 V/2.5 V
operating temperature	0–55 °C

5. Results and Discussion

The estimation of SOC and the identification of battery parameters were verified under two working conditions (US06 and BJDST), at three different temperatures.

5.1. Results of US06

The results of US06 are plotted in Figures 5 and 6. Figure 5a,c,e plots the voltage of the UKF–EKF and the real voltage at 0, 25, and 45 °C, respectively. Figure 5b,d,f correspondingly compares online estimated voltage errors and offline estimated voltage errors at different temperatures; the online estimation is implemented by the UKF–EKF. Figure 6a,c,e describes the results of SOC estimation between the UKF–EKF and UKF at 0, 25, and 45 °C, respectively. It should be noted that at various temperatures, the initial values of the real SOC are 0.8, but the initial values of the SOC for algorithms are 0.6. Figure 6b,d,f shows the identification of R_0 at different temperatures.

It can be seen from Figure 5 and Table 2, in terms of voltage prediction, the root mean square errors (RMSEs) of voltage for online estimation and offline estimation at 0 °C are 17.6 and 34.1 mV, respectively, and they are 6.2 and 17.4 mV for online estimation and offline estimation at 25 °C, while the online and offline RMSEs are 16.9 and 23.3 mV at 45 °C, respectively.

According to Figure 6 and Table 2, for the identification of R_0 at 0 °C, the RMSEs of R_0 for the UKF–EKF is 12.1 mΩ, and it is 5.6 mΩ at 25 °C, while the figure for RMSE of R_0 at 45 °C is 7.9 mΩ. In terms of the SOC estimation, it is clear that at 0 °C, the RMSEs of the SOC for the UKF–EKF and UKF are 1.00% and 2.12%, respectively, and at 25 °C, they are 0.76% and 1.72% for the UKF–EKF and UKF, respectively, while the figures for RMSEs of SOC for the UKF–EKF and UKF are 0.51% and 1.31% at 45 °C, respectively.

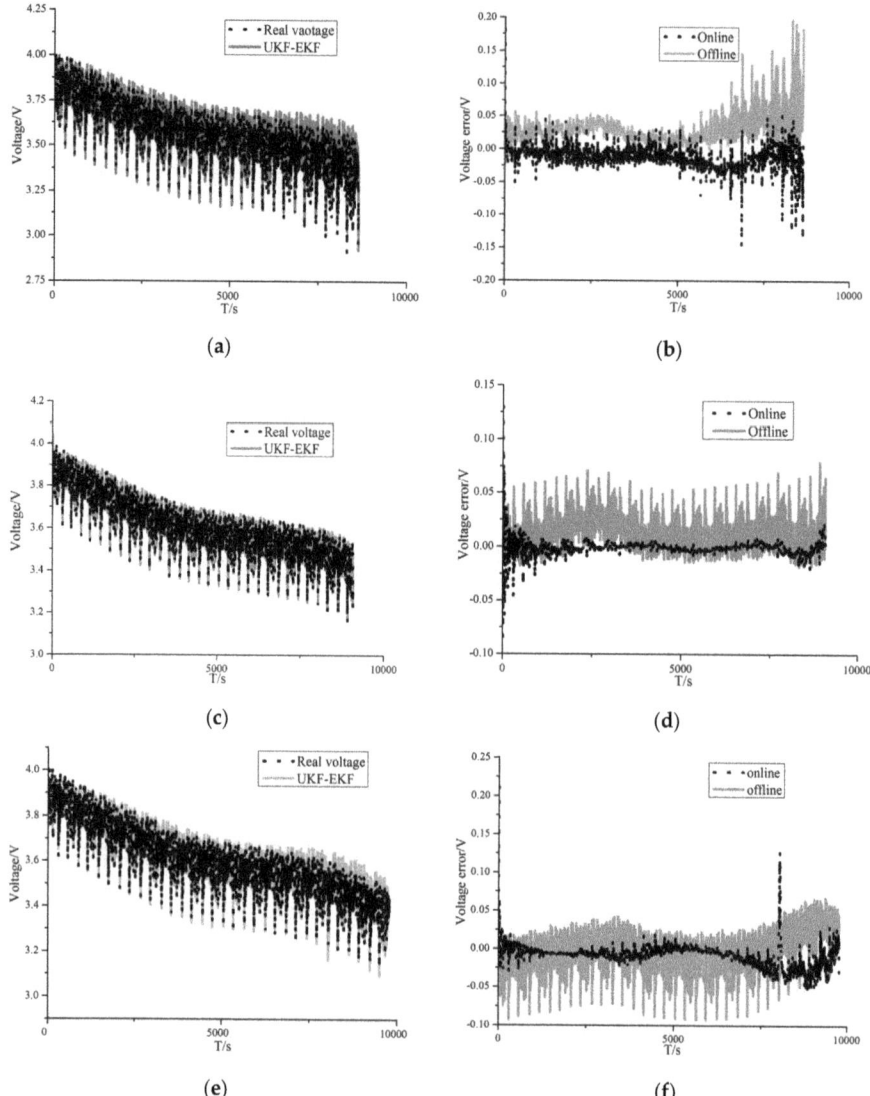

Figure 5. Results of voltage comparison in US06: (**a**) comparison between the UKF–EKF model voltage and real voltage at 0 °C; (**b**) comparison of voltage errors between online and offline estimations at 0 °C; (**c**) comparison between the UKF–EKF model voltage and real voltage at 25 °C; (**d**) comparison of voltage errors between online and offline estimations at 25 °C; (**e**) comparison between the UKF–EKF model voltage and real voltage at 45 °C; (**f**) comparison of voltage errors between online and offline estimations at 45 °C.

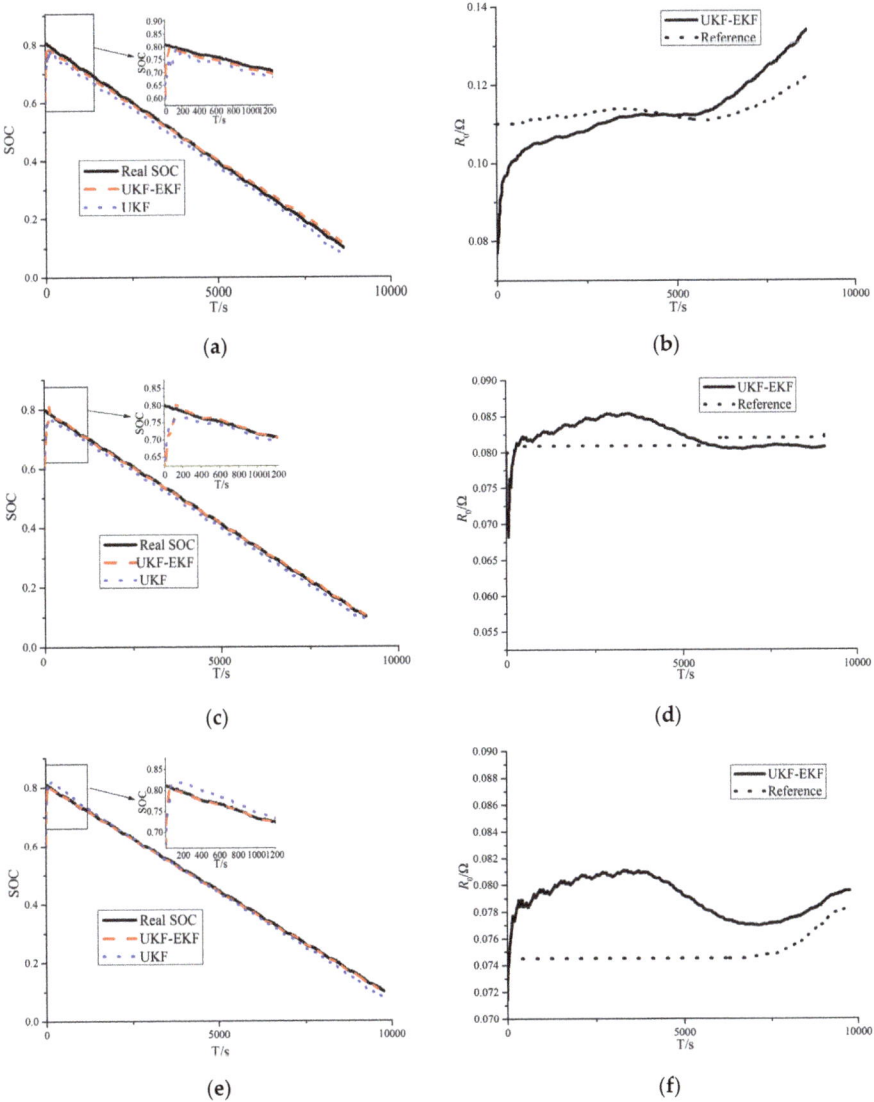

Figure 6. Results of the estimation of the SOC and ohmic resistance (R_0) in US06: (**a**) SOC comparison between the UKF–EKF and UKF at 0 °C; (**b**) R_0 identification of the UKF–EKF at 0 °C; (**c**) SOC comparison between the UKF–EKF and UKF at 25 °C; (**d**) R_0 identification of the UKF–EKF at 25 °C; (**e**) SOC comparison between the UKF–EKF and UKF at 45 °C; (**f**) R_0 identification of the UKF–EKF at 45 °C.

Table 2. The estimation of root mean square errors (RMSEs) for the UKF–EKF and UKF under US06.

Temperatures	0 °C		25 °C		45 °C	
	UKF–EKF	UKF	UKF–EKF	UKF	UKF–EKF	UKF
SOC (%)	1.00	2.12	0.76	1.72	0.51	1.31
Voltage (mV)	17.6	34.1	6.2	17.4	16.9	23.3
R_0 (mΩ)	12.1	-	5.6	-	7.9	-

The results show that the UKF–EKF, which can identify battery parameters online, has a better performance than the UKF in SOC estimation at different temperatures; in addition, the voltage errors of online estimation are smaller. It is clear that at 0 °C, the differences between estimations of the SOC and voltage become greater. Moreover, at 25 and 45 °C, the UKF–EKF has greater accuracy in estimating R_0 than at 0 °C.

5.2. Results of BJDST

The results of BJDST are plotted in Figures 7 and 8. Figure 7a,c,e plots the voltage of the UKF–EKF and the real voltage at 0, 25, and 45 °C, respectively. Figure 7b,d,f correspondingly compares online estimated voltage errors and offline estimated voltage errors at different temperatures; the online estimation is implemented by the UKF–EKF. Figure 8a,c,e describes the results of the SOC estimation between the UKF–EKF and UKF at 0, 25, and 45 °C, respectively. It should be noted that at various temperatures, the initial values of the real SOC are 0.8, but the initial values of the SOC for the algorithms are 0.6. Figure 8b,d,f shows the identification of R_0 at different temperatures.

It can be seen from Figure 7 and Table 3, in terms of voltage prediction, the root mean square errors (RMSEs) of voltage for online estimation and offline estimation at 0 °C are 10.7 and 17.4 mV, respectively, and 5.8 and 9.1 mV for online estimation and offline estimation at 25 °C, while the online and offline RMSEs are 10.1 and 24.7 mV at 45 °C, respectively.

Table 3. The estimation of RMSEs for the UKF–EKF and UKF under BJDST.

Temperatures	0 °C		25 °C		45 °C	
	UKF–EKF	UKF	UKF–EKF	UKF	UKF–EKF	UKF
SOC (%)	0.97	1.95	0.61	1.31	0.82	1.75
Voltage (mV)	10.7	17.4	5.8	9.1	10.1	24.7
R_0 (mΩ)	7.3	-	3.3	-	6.1	-

According to Figure 8 and Table 3, for the identification of R_0 at 0 °C, the RMSE of R_0 for the UKF–EKF is 7.3 mΩ, and it is 3.3 mΩ at 25 °C, while the figure for RMSE of R_0 at 45 °C is 6.1 mΩ. In terms of SOC estimation, it is clear that at 0 °C, the RMSEs of SOC for the UKF–EKF and UKF are 0.97% and 1.95%, respectively, and at 25 °C, they are 0.61% and 1.31% for the UKF–EKF and UKF, respectively, while the figures for RMSEs of SOC for the UKF–EKF and UKF are 0.82% and 1.75% at 45 °C, respectively.

The results showed that the UKF—EKF, which can identify battery parameters online, has a better performance than the UKF in SOC estimation at different temperatures, in addition to the voltage errors of online estimation being smaller. It is clear that at 0 °C, the differences in estimation of SOC and voltage become greater. Moreover, at 25 and 45 °C, the UKF–EKF has greater accuracy in estimating R_0 than at 0 °C.

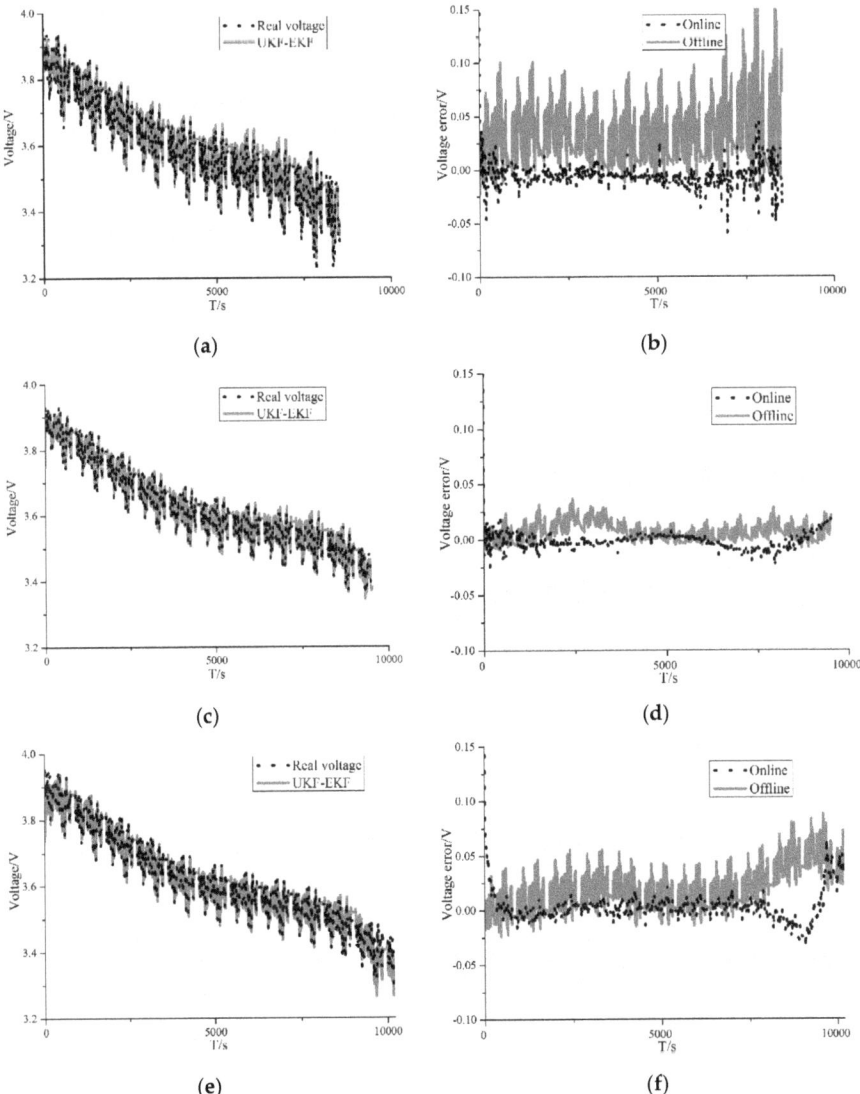

Figure 7. Results of voltage comparisons in BJDST: (**a**) comparison between the UKF–EKF model voltage and real voltage at 0 °C; (**b**) comparison of voltage errors between online and offline estimations at 0 °C; (**c**) comparison between the UKF–EKF model voltage and real voltage at 25 °C; (**d**) comparison of voltage errors between online and offline estimations at 25 °C; (**e**) comparison between the UKF–EKF model voltage and real voltage at 45 °C; (**f**) comparison of voltage errors between online and offline estimations at 45 °C.

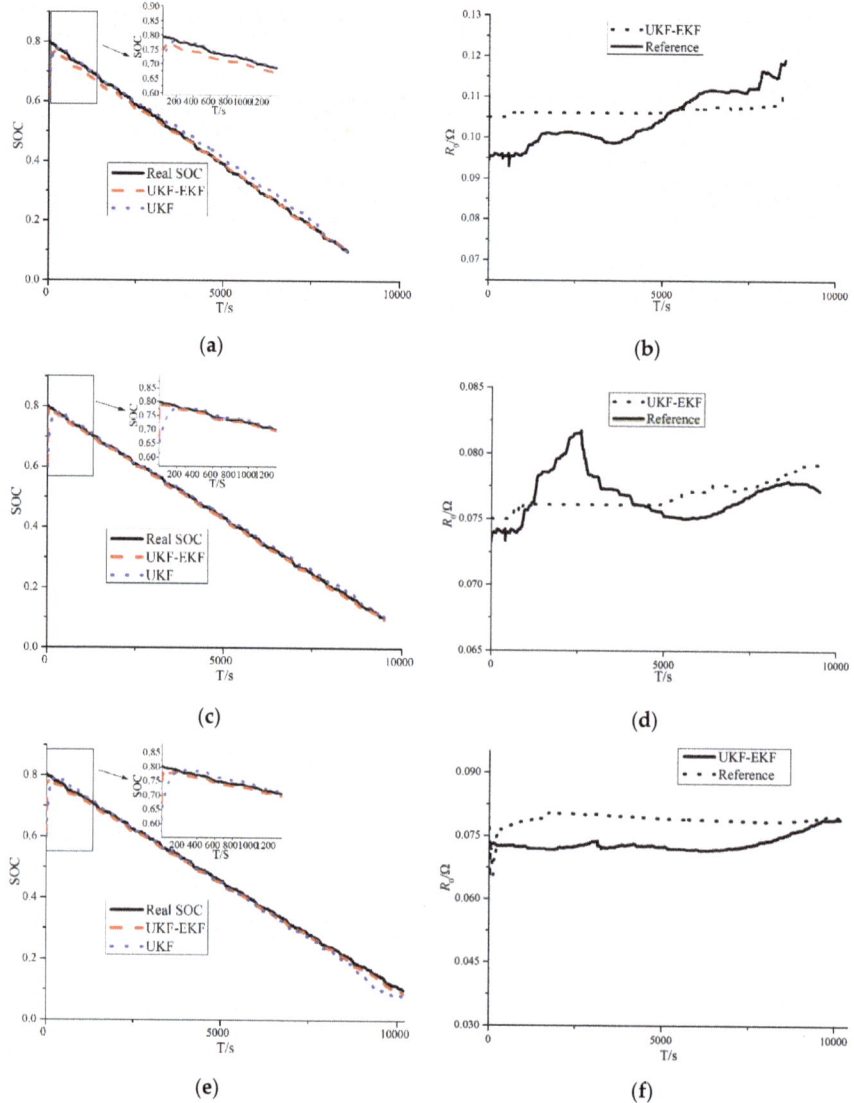

Figure 8. Results of estimation of the SOC and R_0 in BJDST: (**a**) SOC comparison between the UKF–EKF and UKF at 0 °C; (**b**) R_0 identification of the UKF–EKF at 0 °C; (**c**) SOC comparison between the UKF–EKF and UKF at 25 °C; (**d**) R_0 identification of the UKF–EKF at 25 °C; (**e**) SOC comparison between the UKF–EKF and UKF at 45 °C; (**f**) R_0 identification of the UKF–EKF at 45 °C.

In summary, the UKF–EKF established in this study has an accurate and reliable ability to estimate the SOC and identify parameters of the battery at different temperatures and working conditions. As can be seen in Tables 2 and 3, the errors of the SOC estimated by the UKF–EKF are within 1.00%, and the estimated errors of internal resistance are within 15 mΩ. In addition, with the ability to predict internal resistance in real time, the UKF–EKF can provide a better estimation of voltage, with the estimated errors of voltage within 20 mV. It is noted that at the same temperature, the predictions of the SOC, the parameters, and the voltage by the UKF–EKF in the BJDST are better than those in the

US06, mainly because the load current under the US06 condition is more complex than that under the BJDST condition. Hence, the RMSEs of the SOC and voltage in the BJDST condition are smaller than the figures for the US06 condition. In addition, the more complex working conditions influence the estimations of internal resistance; thus, the identification of R_0 in the BJDST condition shows better results. According to the results, battery parameters are temperature-sensitive, because under the same working condition, the predictions of the SOC, voltage, and internal resistance are better at 25 and 45 °C. One of the reasons for this is that at low temperatures (0 °C), the battery model is not accurate enough, which can be reflected in the dynamic hysteresis of OCV in the OCV–SOC curves. At 25 °C, better results of voltage prediction are obtained with smaller errors of parameter estimation than those at 45 °C, which indicates that the accuracy of the battery model at high temperatures (45 °C) is lower than that at room temperature (25 °C). Therefore, compared with other temperatures, the battery model at room temperature is the least different from the real battery, which results in a better SOC estimation at room temperature. Although battery parameters change at different temperatures and in different working conditions, which affects the accuracy of the battery model, the UKF–EKF, with the ability to update battery parameters in real time, reduces external influences through self-correction, and obtains satisfactory SOC estimations.

6. Conclusions

SOC estimation is an important factor for BMS in EVs, as it can provide a basis for the energy management of EVs. In order to obtain an accurate SOC estimation under various conditions, a dual filter to estimate SOC was established.

(1) SOC estimation is greatly influenced by temperature and working conditions, which can be reflected by the differences in the OCV–SOC curves at different temperatures and the prediction results of the two working conditions.

(2) The second-order ECM proposed in this study shows its great performance, since it can accurately reflect the dynamic changes of batteries and the voltage errors of online and offline estimations are within 35 mV.

(3) The UKF–EKF, with its ability to identify battery parameters online, is an improvement on the UKF for SOC estimation, which can be verified in different working conditions (US06 and BJDST) and at different temperatures (0, 25, and 45 °C). Future work will be to verify the feasibility of SOC estimations of other types of batteries and their feasibility at a wider temperature range.

Author Contributions: Conceptualization, F.H. and Y.Z.; Methodology, Y.Z.; Software, Y.Z.; Validation, W.W., Y.Z.; Formal Analysis, Y.Z.; Resources, W.W.; Data Curation, W.W.; Writing-Original Draft Preparation, Y.Z.; Writing-Review & Editing, F.H.; Supervision, F.H.

Funding: This work was funded by the Guizhou Province Science and Technology Support Program, grant number [2019]2155.

Conflicts of Interest: The authors declare no conflict of interest.

References

1. Zhang, X.; Wang, Y.; Yang, D.; Chen, Z.H. An on-line estimation of battery pack parameters and state-of-charge using dual filters based on pack model. *Energy* **2016**, *115*, 219–229. [CrossRef]
2. Ephrem, C.; Kollmeyer, P.J.; Matthias, P.; Emadi, A. State of charge estimation of Li-ion batteries using deep neural networks: A machine learning approach. *J. Power Sources* **2018**, *400*, 242–255.
3. He, W.; Williard, N.; Chen, C.; Pecht, M. State of charge estimation for Li-ion batteries using neural network modeling and unscented Kalman filter-based error cancellation. *Int. J. Elec. Power Energy Syst.* **2014**, *62*, 783–791. [CrossRef]
4. Du, J.; Liu, Z.; Wang, Y. State of charge estimation for Li-ion battery based on model from extreme learning machine. *Control Eng. Pract.* **2014**, *26*, 11–19. [CrossRef]
5. Cao, W.P.; Ming, Z.; Wang, X.Z.; Cai, S.B. Improved bidirectional extreme learning machine based on enhanced random searsch. *Memet. Comput.* **2017**, *11*, 19–26. [CrossRef]

6. Hu, J.N.; Hu, J.J.; Lin, H.B.; Li, X.P.; Jiang, C.L.; Qiu, X.H.; Li, W.S. State-of-charge estimation for battery management system using optimized support vector machine for regression. *J. Power Sources* **2014**, *269*, 682–693. [CrossRef]
7. Xiong, B.Y.; Zhao, J.Y.; Wei, Z.B.; Skyllas-Kazacos, M. Extended Kalman filter method for state of charge estimation of vanadium redox flow battery using thermal-dependent electrical model. *J. Power Sources* **2014**, *262*, 50–61. [CrossRef]
8. He, H.G.; Xiong, R.; Fan, J.X. Evaluation of Lithium-Ion Battery Equivalent Circuit Models for State of Charge Estimation by an Experimental Approach. *Energies* **2011**, *4*, 582–598. [CrossRef]
9. He, J.G.; Chen, D.; Pan, C.F.; Chen, L.; Wang, S.H. State of charge estimation of power Li-ion batteries using a hybrid estimation algorithm based on UKF. *Electrochim. Acta* **2016**, *21*, 101–109.
10. Peng, J.K.; Luo, J.Y.; He, H.W.; Lu, B. An improved state of charge estimation method based on cubature Kalman filter for lithium-ion batteries. *Appl. Energy* **2019**, *253*, 113520. [CrossRef]
11. Xiong, R.; Gong, X.Z.; Mi, C.C.; Sun, F.C. A robust state-of-charge estimator for multiple types of lithium-ion batteries using adaptive extended Kalman filter. *J. Power Sources* **2013**, *243*, 805–816. [CrossRef]
12. Peng, S.M.; Chen, C.; Shi, H.B.; Yao, Z.L. State of Charge Estimation of Battery Energy Storage Systems Based on Adaptive Unscented Kalman Filter with a Noise Statistics Estimator. *IEEE Access* **2017**, *5*, 13202–13212. [CrossRef]
13. Xia, B.Z.; Wang, H.Q.; Yong, T.; Wang, M.W. State of Charge Estimation of Lithium-Ion Batteries Using an Adaptive Cubature Kalman Filter. *Energies* **2015**, *8*, 5916–5936. [CrossRef]
14. Xiong, R.; Sun, F.C.; He, H.W.; Nguyen, T.D. A data-driven adaptive state of charge and power capability joint estimator of lithium-ion polymer battery used in electric vehicles. *Energy* **2013**, *63*, 295–308. [CrossRef]
15. Wei, Z.B.; Zhao, J.Y.; Zou, C.F.; Lim, T.M.; Tseng, K.J. Comparative study of methods for integrated model identification and state of charge estimation of lithium-ion battery. *J. Power Sources* **2018**, *402*, 189–197. [CrossRef]
16. Shen, P.; Ouyang, M.; Lu, L.; Li, J.; Feng, X. The co-estimation of state of charge, state of health, and state of function for lithium-ion batteries in electric vehicles. *IEEE. Trans. Veh. Technol.* **2018**, *67*, 92–103. [CrossRef]
17. Sepasi, S.; Ghorbani, R.; Liaw, B.Y. Improved extended Kalman filter for state of charge estimation of battery pack. *J. Power Sources* **2014**, *255*, 368–376. [CrossRef]
18. Shrivastava, P.; Soon, T.K.; Idris, M.Y.I.B.; Mekhilef, S. Overview of model-based online state-of-charge estimation using Kalman filter family for lithium-ion batteries. *Renew. Sust. Energy Rev.* **2019**, *113*, 109–233. [CrossRef]
19. Zheng, F.D.; Xing, Y.J.; Jiang, J.C.; Sun, B.X.; Kim, J.H. Influence of different open circuit voltage tests on state of charge online estimation for lithium-ion batteries. *Appl. Energy* **2016**, *183*, 513–525. [CrossRef]

© 2019 by the authors. Licensee MDPI, Basel, Switzerland. This article is an open access article distributed under the terms and conditions of the Creative Commons Attribution (CC BY) license (http://creativecommons.org/licenses/by/4.0/).

Article

Estimation of the State of Charge of Lithium Batteries Based on Adaptive Unscented Kalman Filter Algorithm

Jiechao Lv, Baochen Jiang, Xiaoli Wang *, Yirong Liu and Yucheng Fu

School of Mechanical, Electrical and Information Engineering, Shandong University, Weihai 264209, China; lvjiechao@mail.sdu.edu.cn (J.L.); jbc@sdu.edu.cn (B.J.); lyr@mail.sdu.edu.cn (Y.L.); whj3105@126.com (Y.F.)
* Correspondence: wxl@sdu.edu.cn; Tel.: +86-138-6302-6640

Received: 24 July 2020; Accepted: 31 August 2020; Published: 2 September 2020

Abstract: The state of charge (SOC) estimation of the battery is one of the important functions of the battery management system of the electric vehicle, and the accurate SOC estimation is of great significance to the safe operation of the electric vehicle and the service life of the battery. Among the existing SOC estimation methods, the unscented Kalman filter (UKF) algorithm is widely used for SOC estimation due to its lossless transformation and high estimation accuracy. However, the traditional UKF algorithm is greatly affected by system noise and observation noise during SOC estimation. Therefore, we took the lithium cobalt oxide battery as the analysis object, and designed an adaptive unscented Kalman filter (AUKF) algorithm based on innovation and residuals to estimate SOC. Firstly, the second-order RC equivalent circuit model was established according to the physical characteristics of the battery, and the least square method was used to identify the parameters of the model and verify the model accuracy. Then, the AUKF algorithm was used for SOC estimation; the AUKF algorithm monitors the changes of innovation and residual in the filter and updates system noise covariance and observation noise covariance in real time using innovation and residual, so as to adjust the gain of the filter and realize the optimal estimation. Finally came the error comparison analysis of the estimation results of the UKF algorithm and AUKF algorithm; the results prove that the accuracy of the AUKF algorithm is 2.6% better than that of UKF algorithm.

Keywords: SOC; second-order RC equivalent circuit model; system noise covariance; observation noise covariance; AUKF

1. Introduction

In recent years, with the escalating energy crisis and environmental problems, low-pollution, high-efficiency electric vehicles (EVs) have become a hot spot in the automotive industry. Lithium-ion batteries have the characteristics of small size, light weight, high energy density, large output power and high safety performance, and have become the first choice for energy storage devices of EVs [1–3]. State of charge (SOC) is used to directly reflect the remaining capacity of the battery, which is an important basis for the vehicle control system to formulate an optimal energy management strategy. SOC is an important battery performance parameter; accurate estimation of SOC is of great significance to improve battery safety performance, extend battery life and ensure reliable operation of battery system [4,5].

At present, the commonly used SOC estimation methods for lithium batteries include the ampere-hour integration method, the open circuit voltage method, the neural network method, the particle filter algorithm and the Kalman Filter (KF) method. Among them, the ampere-hour integration method estimates the SOC of the battery by accumulating the amounts of charge and discharge, and at the same time compensates the estimated SOC according to the self-discharge rate [6,7]. The ampere-hour integration method is relatively simple; it can dynamically estimate the battery SOC, but the current integration needs to obtain the initial SOC value, and the battery current must be accurately collected, which leads to the accumulation of SOC estimation errors over time. In practical applications, the ampere-hour integration method is usually used in combination with other methods to improve the estimation accuracy.

The open circuit voltage method is to indirectly fit the corresponding relationship between the open circuit voltage and the battery SOC, according to the relationship between the open circuit voltage of the battery and the lithium ion concentration in the battery [8,9]. The open circuit voltage method requires the battery to be placed statically for a long time to obtain a stable terminal voltage. Therefore, the open circuit voltage method cannot be used to estimate the SOC of the battery online in real time.

The neural network method is an algorithm for simulating the human brain and its neurons to deal with nonlinear systems, without in-depth study of the internal structure of the battery. The neural network method only needs to extract a large number of input and output samples from the target battery in accordance with its working characteristics in advance, and input it into the system established by using this method, and the SOC of the battery can be obtained [10,11]. The neural network method has high operational complexity, and it needs to extract a large amount of comprehensive target sample data to train the system. The input training data and training method will affect the accuracy of SOC estimation to a large extent.

The particle filtering is a process of approximating the probability density function by finding a set of random samples propagating in the state space, replacing the integral operation with the sample mean and then obtaining the minimum variance estimation process of the system state [12,13]. The particle filter algorithm is suitable for nonlinear non-Gaussian systems. The more particles used, the more accurate the SOC estimation value. However, as the number of particles increases, the calculation load increases. Additionally, particle degradation and insufficient particle diversity will seriously affect the SOC estimation results.

The KF algorithm is a type of optimized autoregressive data filtering algorithm. The essence of the algorithm is to make an optimal estimate of complex dynamic systems according to the principle of least mean square error [14–18]. KF algorithm overcomes the serious shortcoming of the current integration dependence on the initial value, and does not require a large number of sample data, and can be used to estimate the battery SOC online. In the SOC estimation of electric vehicle power batteries with complex operating conditions, the KF algorithm has a significant application value, and has become a hot spot in the research of battery SOC estimation algorithms in recent years [19,20]. The KF is an algorithm that uses the linear system state equation to observe the system input and output data to optimally estimate the state of the system.

Since KF cannot solve the problem of nonlinear systems, study [21] used the extended Kalman filter (EKF) to expand nonlinear systems into linear systems using Taylor series. EKF is an extended form of the standard Kalman filter in non-linear situations, and it is a highly efficient recursive filter. The basic idea of EKF is to use Taylor series expansion to linearize the nonlinear system, and then use the Kalman filter framework to filter the signal, so it is a sub-optimal filter. EKF algorithm is used for SOC estimation in battery management systems (BMSs), and has achieved good results in SOC estimation based on equivalent circuit model [22–24]. Although this method solves the nonlinear problem, it ignores high-order terms and increases linear errors, which may cause the filter to diverge.

Reference [25] used the unscented Kalman filter (UKF) to perform an unscented transformation on a nonlinear system without ignoring higher-order terms, which improved the accuracy of the estimation. UKF is a combination of unscented transform and standard Kalman filter system. Through unscented transform, the nonlinear system equation is suitable for the standard Kalman system under the linear assumption. The basic idea of UKF is Kalman filtering and unscented transform, which can effectively overcome the problems of low accuracy and poor stability of EKF estimation. As high-order terms are not ignored, the calculation accuracy of nonlinear distribution statistics is high. However, the uncertainty of the battery model and system noise is not considered. Uncertainty of model noise and system noise will lead to increased error, slow convergence speed and filter divergence. Reference [26] introduces adaptive filtering on the basis of UKF, and replaces system noise covariance and observation noise covariance of UKF with adaptive-filter-estimated system noise covariance and observation noise covariance, respectively. In order to update the system error and the observation error in real time, the filtering effect is relatively good, but the adaptive filtering cannot truly reflect the system noise and the observation noise error, so it can be further improved.

In the traditional UKF algorithm [27,28], the system noise covariance and the observation noise covariance are usually set as constants, which cannot truly reflect the dynamic characteristics of noise, and have a certain influence on the accuracy of SOC estimation. In view of the shortcomings of the traditional UKF algorithm in the case of low model accuracy and uncertain noise, we designed an adaptive unscented Kalman filter (AUKF). The AUKF algorithm monitors the dynamic changes of innovation and residual in the filter in real time; corrects the system noise covariance and observation noise covariance in real time; and adjusts the filter gain to improve the estimation accuracy.

The organizational structure of this paper is as follows. In Section 1, the common methods for battery SOC estimation are introduced, and the methods designed in this paper are briefly introduced. In Section 2, the second-order RC equivalent circuit model is established, the parameters are identified and the accuracy of the model is verified. In Section 3, the traditional UKF algorithm and the AUKF algorithm designed in this paper are introduced. In Section 4, the convergence speed and estimation accuracy of the UKF algorithm and AUKF algorithm are compared through experiments. In Section 5, the work and research results of this paper are summarized.

2. Lithium Battery Model

2.1. The Second-Order RC Equivalent Circuit Model of a Lithium-Ion Battery

An accurate battery model is the basis for SOC estimation. Battery models can be roughly divided into electrochemical models [29,30], mathematical models [31] and equivalent circuit models [32,33]. Although the accuracy of the electrochemical model is high, the structure is complex and difficult to implement, and it is not suitable for modeling actual working conditions. The mathematical model has a simple structure and is easy to calculate, but it is difficult to describe the external characteristics of the battery. Considering the complexity and accuracy of the battery model, this paper uses a second-order RC equivalent circuit model [34]; the schematic diagram is shown in Figure 1.

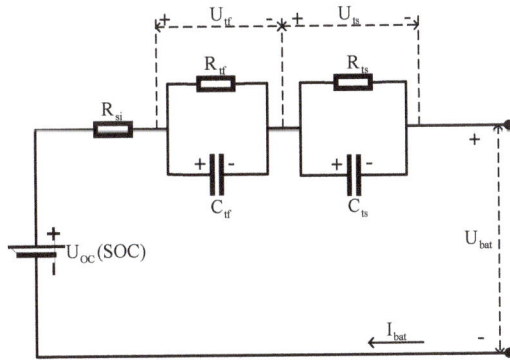

Figure 1. Schematic diagram of the second-order RC equivalent circuit.

In Figure 1, $U_{OC}(SOC)$ represents the battery open circuit voltage related to SOC; I_{bat} represents the open circuit current of the battery, and the discharge current is a positive value; U_{bat} represents the battery terminal voltage; R_{si} represents the ohmic internal resistance of the battery; R_{tf} and C_{tf} represent the polarization resistance and polarization capacitance of the battery respectively; R_{ts} and C_{ts}

represent concentration polarization resistance and concentration polarization capacitance respectively; U_{tf} and U_{ts} represent the voltage across the polarization capacitance and the concentration polarization capacitance, respectively. According to Kirchhoff's law, the state equation and output equation of the equivalent circuit can be obtained:

$$\begin{cases} \frac{dU_{tf}}{dt} = -\frac{U_{tf}}{R_{tf}C_{tf}} + \frac{I_{bat}}{C_{tf}} \\ \frac{dU_{ts}}{dt} = -\frac{U_{ts}}{R_{ts}C_{ts}} + \frac{I_{bat}}{C_{ts}} \\ \frac{dSOC}{dt} = -\frac{I_{bat}}{Q_{bat}} \end{cases} \quad (1)$$

$$U_{bat} = U_{OC}(SOC) - R_{si}I_{bat} - U_{tf} - U_{ts} \quad (2)$$

where Q_{bat} is the rated capacity of the battery.

2.2. Parameter Identification

Parameter identification technology is a technology that combines theoretical models and experimental data for prediction. Parameter identification determines the parameter values of a group of models based on the model established by the experimental data, so that the numerical results calculated by the model can better fit the test data, so that the unknown process can be predicted.

In this section, we identify the parameters through the voltage response curve of battery discharge and combine Equations (1) and (2); the parameters to be identified are $R_{si}, R_{tf}, C_{tf}, R_{ts}, C_{ts}$ and function relationship $U_{OC}(SOC)$.

The cell model used in the experiment in this paper was the SAMSUNG 30Q INR18650 power lithium cell. The specific parameters of the cell are shown in Table 1. The experimental object of this paper is a battery composed of 10 parallel lithium cobalt oxide cells. The battery was discharged by 1 C pulsed for 3 min, placed statically for 2 h and discharged to the cut-off voltage in cycles. The pulsed discharge voltage is shown in Figure 2a, and the pulsed discharge current is shown in Figure 2b.

Table 1. Power lithium cell parameters.

Parameter	Value
Cell model	SAMSUNG 30Q INR18650
Rated capacity	3000 mA h
Rated voltage	3.6 V
Discharge cut-off voltage	2.5 V
Weight	48.1 ± 1.5 g
Size	18.2 mm (D) × 65.0 mm (H)

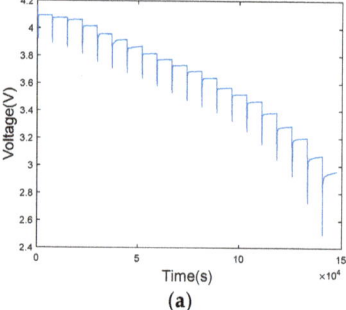
(a)

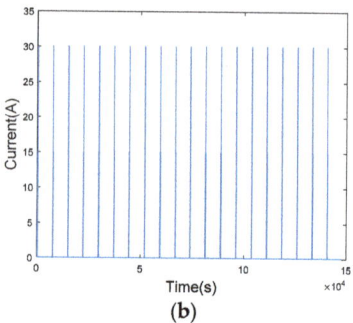
(b)

Figure 2. Pulsed discharge current and voltage of the battery. (a) Battery pulsed discharge voltage. (b) Battery pulsed discharge current.

2.2.1. Parameter Identification of the Functional Relationship between U_{oc} and SOC

The battery SOC and open circuit voltage U_{oc} were obtained by the static method [35], and the corresponding values of SOC and open circuit voltage U_{oc} are shown in Table 2.

Table 2. U_{oc} and SOC corresponding relationship value.

U_{OC} (V)	SOC	U_{OC} (V)	SOC
4.1617	1	3.7317	0.5034
4.0913	0.9503	3.6892	0.4537
4.0749	0.9007	3.6396	0.4040
4.0606	0.8510	3.5677	0.3543
4.0153	0.8013	3.5208	0.3046
3.9592	0.7517	3.4712	0.2550
3.9164	0.7020	3.3860	0.2053
3.8687	0.6524	3.2880	0.1556
3.8163	0.6027	3.2037	0.1059
3.7735	0.5530	3.0747	0.0563

Use MATLAB to perform the least squares fitting of the data in Table 2 to obtain the equation of the functional relationship between U_{oc} and SOC; the equation is shown in Equation (3). The fitted relationship curve between U_{oc} and SOC is shown in Figure 3.

$$\begin{aligned} U_{OC}(SOC) = & 122.4786 * SOC^8 - 401.4734 * SOC^7 + 485.6818 * SOC^6 \\ & -239.2806 * SOC^5 + 3.7304 * SOC^4 + 44.9020 * SOC^3 - 19.8057 * SOC^2 \\ & +5.0932 * SOC + 2.8341 \end{aligned} \quad (3)$$

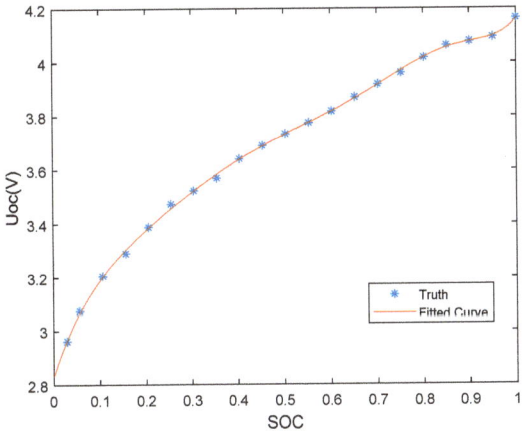

Figure 3. U_{oc}–SOC fitting curve.

2.2.2. Parameter Identification of Resistance and Capacitance

This paper combines the characteristics of resistance and capacitance, and analyzes the voltage response curve of the battery pulsed discharge to identify the resistance and capacitance. The partial discharge voltage diagram of the battery pulsed discharge is shown in Figure 4, and the battery voltage response curve can be divided into four stages:

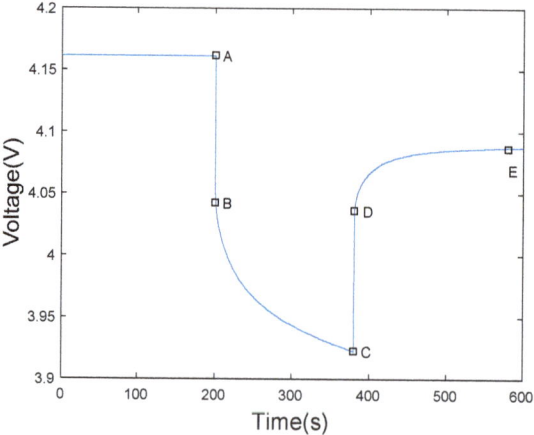

Figure 4. Partial enlargement of pulsed discharge voltage.

Section A-B: The battery turns from a static state to a discharged state, and the terminal voltage drops abruptly. From the second-order equivalent circuit diagram, it can be seen that U_{tf} and U_{ts} cannot be abruptly changed. The sudden drop in the voltage of section A-B is caused by the ohmic internal resistance R_{si}.

Section B-C: During the continuous discharge, electrochemical polarization and concentration polarization work together to make the voltage drop in the form of exponential changes. Before section B-C, U_{tf} and U_{ts} are zero, so section B-C can be regarded as a zero state response.

Section C-D: The battery discharge current disappears and the battery terminal voltage rises rapidly. It is the same as the section A-B. It can be considered that it is caused by the ohmic internal resistance R_{si}.

Section D-E: The battery is at rest. Due to the electrochemical polarization and concentration difference, the voltage is slowly increased. At this time, there is no current discharge, which can be regarded as zero input response.

According to the sections A-B and C-D in Figure 4, the ohmic internal resistance can be obtained:

$$R_{si} = \frac{(U_A - U_B) + (U_D - U_C)}{2I_{bat}} \qquad (4)$$

where U_A, U_B, U_C, U_D are the battery terminal voltages corresponding to points A, B, C and D in Figure 4 respectively; I_{bat} is the discharge current of the battery.

Solving the differential equation according to Equation (1) gives Equation (5):

$$\begin{cases} U_{tf}(t) = U_{tf}(0)e^{-t/\tau_{tf}} + I_{bat}R_{tf}(1 - e^{-t/\tau_{tf}}) \\ U_{ts}(t) = U_{ts}(0)e^{-t/\tau_{ts}} + I_{bat}R_{ts}(1 - e^{-t/\tau_{ts}}) \end{cases} \qquad (5)$$

where $\tau_{tf} = R_{tf}C_{tf}, \tau_{ts} = R_{ts}C_{ts}$ are the fast time constant and slow time constant respectively; $U_{tf}(0)$ and $U_{ts}(0)$ are the initial voltages across C_{tf}, C_{ts}, respectively.

According to Figure 4, the discharge current is zero in the DE segment, as a zero input response state. Taking point D as the starting moment, the zero input response expression of the RC loop can be obtained as shown in Equation (6):

$$\begin{cases} U_{tf} = U_{tf}(0)e^{-t/\tau_{tf}} \\ U_{ts} = U_{ts}(0)e^{-t/\tau_{ts}} \end{cases} \qquad (6)$$

Combined with Equation (2), the battery output equation at zero input response is:

$$U_{bat}(t) = U_{OC}(SOC) - U_{tf}(0)e^{-t/\tau_{tf}} - U_{ts}(0)e^{-t/\tau_{ts}} \qquad (7)$$

Equation (7) can be simplified:

$$U_{bat}(t) = U_{OC}(SOC) - b_1 e^{-\lambda_1 t} - b_2 e^{-\lambda_2 t} \qquad (8)$$

where $\tau_{tf} = \frac{1}{\lambda_1}, \tau_{ts} = \frac{1}{\lambda_2}, U_{tf}(0) = b_1, U_{ts}(0) = b_2$.

By using Equation (8) as the fitting function, and using MATLAB to perform the least squares fitting on the DE segment in Figure 4, the value of $b_1, b_2, \lambda_1, \lambda_2$ can be obtained.

According to the BC segment in Figure 4, it can be regarded as a zero state response. Taking point B as the initial moment, the expression of the zero state response of the RC loop can be obtained as shown in Equation (9):

$$\begin{cases} U_{tf}(t) = I_{bat} R_{tf}(1 - e^{-t/\tau_{tf}}) \\ U_{ts}(t) = I_{bat} R_{ts}(1 - e^{-t/\tau_{ts}}) \end{cases} \qquad (9)$$

Combined with Equation (2), the battery output equation at zero state response is:

$$U_{bat}(t) = U_{OC}(SOC) - I_{bat} R_{si} - I_{bat} R_{tf}(1 - e^{-t/\tau_{tf}}) - I_{bat} R_{ts}(1 - e^{-t/\tau_{ts}}) \qquad (10)$$

Take τ_{tf}, τ_{ts} obtained by the Equation (8) fitting into Equation (10), use Equation (10) as the fitting function and use MATLAB to perform the least squares fitting of the BC segment in Figure 4; that will provide the values of a_1, a_2, and then the value of R_{tf}, R_{ts} is obtained:

$$\begin{cases} R_{tf} = \frac{a_1}{I_{bat}} \\ R_{ts} = \frac{a_2}{I_{bat}} \end{cases} \qquad (11)$$

According to $\tau_{tf} = R_{tf} C_{tf}, \tau_{ts} = R_{ts} C_{ts}$, the value of C_{tf}, C_{ts} can be obtained.

According to the battery discharge voltage curve and battery characteristics, the results of identifying the parameters of the battery model by using the least square method in MATLAB are shown in Figure 5.

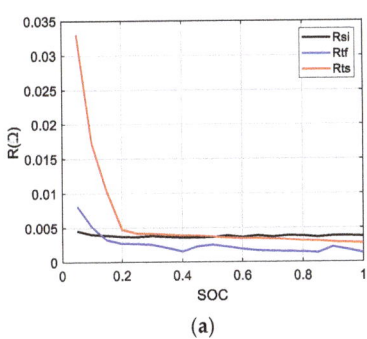
(a)

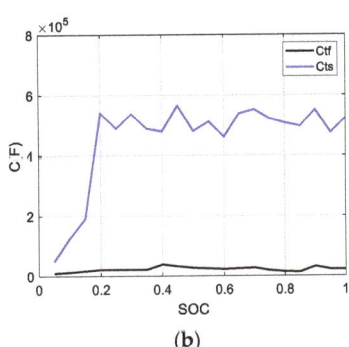
(b)

Figure 5. Result of parameter identification. (a) The value of the resistance. (b) The value of the capacitor.

Figure 5 shows the change of battery resistance and capacitance with SOC when the battery is discharged with constant current pulsed at a constant temperature of 25 °C. When the battery SOC value is between 0% and 20%, the resistance and capacitance values change greatly, and when the battery SOC value is between 20% and 100%, the resistance and capacitance values change relatively little. Considering that the lower limit SOC value of battery in the actual working environment is 20%, this paper takes the average value of the resistance and capacitance of the battery's SOC in the range of 20–100% as the battery parameters for the subsequent SOC estimation experiment.

The average value of the resistance and capacitance of the battery with SOC in the range of 20–100% is shown in Table 3.

Table 3. Lithium-ion battery parameter identification results.

R_{si} (Ω)	R_{tf} (Ω)	R_{ts} (Ω)	C_{tf} (F)	C_{ts} (F)
0.0037	0.0019	0.0035	23,340	501,270

2.2.3. Verifying the Battery Model

Take the battery parameters identified in Table 3 into Equations (1) and (2), use the pulsed discharge current as input and compare the output terminal voltage with the actual terminal voltage. The comparison between the true value of the battery terminal voltage and the model value is shown in Figure 6; the model error value of the battery terminal voltage is shown in Figure 7; and the relevant parameters of the model error are shown in Table 4.

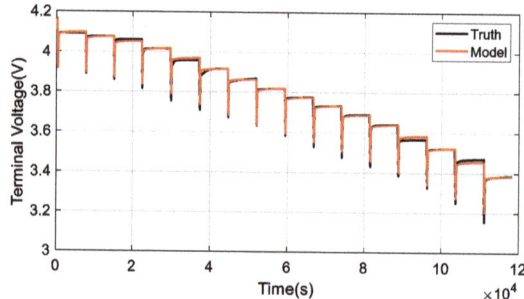

Figure 6. True value and modeled values of terminal voltage.

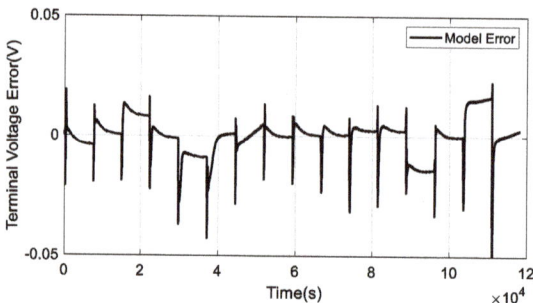

Figure 7. Terminal voltage error.

Table 4. Model error parameters.

Error Type	MAE	RMSE
Value	0.51%	0.8%

In Figure 6, the actual value of the battery terminal voltage is compared with the model value of the battery terminal voltage; the terminal voltage curve of the battery model is basically consistent with the true terminal voltage curve of the battery. In Figure 7, the terminal voltage error of the battery model is shown; the terminal voltage error value of the battery model fluctuates around ±0.05 V. In Table 4, the terminal voltage error value of the battery model is calculated; the mean absolute error (MAE) of the terminal voltage of the battery model is 0.51%; the root mean square error (RMSE) of the terminal voltage of the battery model is 0.8%. The above results prove that the second-order RC equivalent circuit model of the battery designed in this paper is reasonable and reliable, and the battery model was able to be used in subsequent experiments.

3. Design of the SOC Estimation Algorithm

For a nonlinear system, the state equation and observation equation considering the system noise and observation noise are as shown in Equation (12):

$$\begin{cases} x_k = F(x_{k-1}, u_k) + w \\ y_k = G(x_{k-1}, u_k) + v \end{cases} \tag{12}$$

where k is the current moment, $F(x_{k-1}, u_k)$ is the nonlinear system state transition equation, $G(x_{k-1}, u_k)$ is the nonlinear observation equation, x_k is the state variable, u_k is the known input, y_k is the observation signal, w is the system noise and v is the observation noise.

According to the second-order equivalent circuit model of the battery, combining Equations (1) and (2), the discretized state equation and observation equation of the equivalent circuit model of the battery can be shown in Equation (13):

$$\begin{cases} \begin{bmatrix} U_{tf}(k) \\ U_{ts}(k) \\ SOC(k) \end{bmatrix} = \begin{bmatrix} e^{-\frac{\Delta t}{\tau_{tf}}} & 0 & 0 \\ 0 & e^{-\frac{\Delta t}{\tau_{ts}}} & 0 \\ 0 & 0 & 1 \end{bmatrix} \begin{bmatrix} U_{tf}(k-1) \\ U_{ts}(k-1) \\ SOC(k-1) \end{bmatrix} + \begin{bmatrix} R_{tf}(1 - e^{-\frac{\Delta t}{\tau_{tf}}}) \\ R_{ts}(1 - e^{-\frac{\Delta t}{\tau_{ts}}}) \\ -\frac{\Delta t}{Q_n} \end{bmatrix} I_{bat}(k) \\ U_{bat}(k) = U_{OC}(SOC) + [-1 \ -1 \ 0] \begin{bmatrix} U_{tf}(k) \\ U_{ts}(k) \\ SOC(k) \end{bmatrix} - I_{bat}(k) R_{si} \end{cases} \tag{13}$$

Equations (13) can be simplified to Equations (14):

$$\begin{cases} x_k = A_k x_{k-1} + B_k u_k \\ y_k = U_{OC}(SOC) + Cx_k - R_{si} u_k \end{cases} \tag{14}$$

where

$$x_k = \begin{bmatrix} U_{lf}(k) \\ U_{ts}(k) \\ SOC(k) \end{bmatrix}, A_k = \begin{bmatrix} e^{-\frac{\Delta t}{\tau_{tf}}} & 0 & 0 \\ 0 & e^{-\frac{\Delta t}{\tau_{ts}}} & 0 \\ 0 & 0 & 1 \end{bmatrix}, B_k = \begin{bmatrix} R_{tf}(1 - e^{-\frac{\Delta t}{\tau_{tf}}}) \\ R_{ts}(1 - e^{-\frac{\Delta t}{\tau_{ts}}}) \\ -\frac{\Delta t}{Q_n} \end{bmatrix}, I_{bat}(k) = u_k, U_{bat}(k) =$$

$y(k), C = [-1 \ -1 \ 0]$, R_{si} is the ohmic internal resistance of the battery.

According to the KF principle, combining Equations (12) and (14), the first derivative of the nonlinear observation equation is calculated at the current state value, and the observation matrix can be obtained as Equation (15).

$$H_k = \frac{\partial G(x_k, u_k)}{\partial x_k} = \begin{bmatrix} -1 & -1 & \frac{\partial U_{oc}(SOC)}{\partial x} \end{bmatrix} \tag{15}$$

3.1. Design of the Unscented Kalman Filter Algorithm

The unscented Kalman filter (UKF) is a combination of the unscented transform (UT) and the standard Kalman filter system, and uses the unscented transform to adapt the nonlinear system equations to the standard Kalman system under the linear assumption. UKF uses statistical linearization technology, which mainly linearizes the nonlinear function of random variables through linear regression of n Sigma points collected in the prior distribution. This linearization is more accurate than Taylor series linearization. The basic idea of UKF is Kalman filtering and unscented transform. Since UKF does not ignore high-order terms, it can effectively overcome the problems of low accuracy and poor stability of EKF estimation.

We conducted simulation experiments in MATLAB, and used the UKF algorithm to perform SOC estimation experiments under urban dynamometer driving schedule (UDDS) conditions, where the initial value $x_0 = [0\ 0\ 0.6]^T$, $P_0 = diag([10^{-5}, 10^{-5}, 10^{-3}])$, $Q = 10^{-7} \times eye(3)$, $R = 1$.

The UDDS operating conditions are shown in Figure 8. The estimation terminal voltage of the battery using the UKF algorithm for SOC estimation is shown in Figure 9. The SOC estimation results of the battery using the UKF algorithm for SOC estimation are shown in Figure 10.

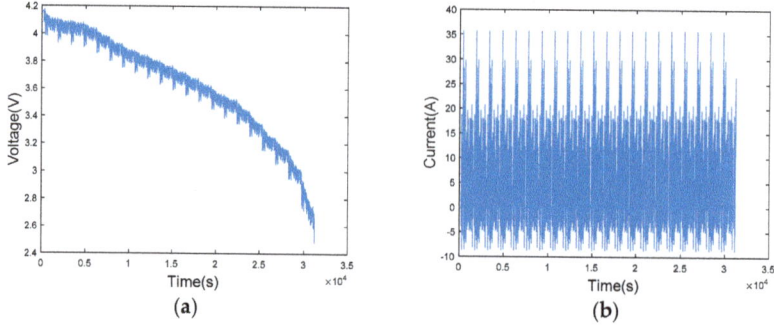

Figure 8. UDDS operating conditions. (**a**) UDDS operating voltage. (**b**) UDDS operating current.

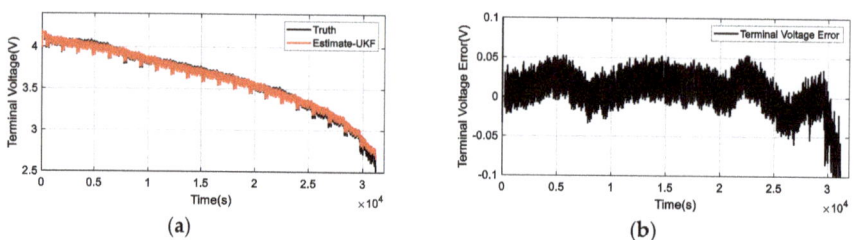

Figure 9. The terminal voltage result of the battery was estimated using the UKF algorithm. (**a**) Estimated value and true value of terminal voltage of the battery. (**b**) Estimated error value of the terminal voltage of the battery.

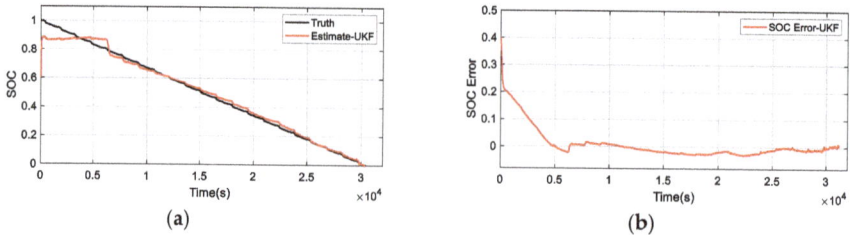

Figure 10. The battery SOC estimation result was estimated using the UKF algorithm. (**a**) Estimated value and true value of SOC of the battery. (**b**) Estimated error value of SOC of the battery.

In Figure 9, the battery terminal voltage estimated by the UKF algorithm is shown. In Figure 9a, the estimated terminal voltage is compared with the true terminal voltage, and the estimated terminal voltage is not much different from the true terminal voltage. In Figure 9b, the terminal voltage error value of the battery fluctuates greatly. In Figure 10, the SOC of the battery estimated by UKF is shown. In Figure 10a, the estimated SOC of the battery differs greatly from the true value before 5000 seconds. In Figure 10b, the SOC estimation error of the battery just approached zero after 5000 seconds, and the convergence speed of the filter is slow.

3.2. Design of the Adaptive Unscented Kalman Filter Algorithm

The UKF algorithm uses UT to replace Taylor series expansion to transform a nonlinear system into a linear system, improving the accuracy of the algorithm. However, in the UKF algorithm, the system model noise and observation noise are set as constants, which cannot reflect the effect of real noise on the filter, which causes the SOC estimation error to increase or even diverge. In order to solve the above problems, we designed an AUKF algorithm; the algorithm is improved on the basis of the UKF algorithm; the algorithm monitors the change of innovation and residual in the filter in real time, and calculates the variance of innovation and residual by the moving window method. The system noise covariance is corrected in real time by the innovation variance, and the observation noise covariance is corrected in real time by the residual variance.

The AUKF algorithm process is as follows:

(1) Determine the initial value of state value $\hat{x}_0$ and the initial value of state error covariance P_0:

$$\hat{x}_0 = E[x_0] \tag{16}$$

$$P_0 = E\left[(x_0 - \hat{x}_0)(x_0 - \hat{x}_0)^T\right] \tag{17}$$

(2) Calculate Sigma point:

$$\begin{cases} x_k^0 = \hat{x}_{k-1} \\ x_{k-1}^i = \hat{x}_{k-1} + \sqrt{(L+\lambda)P_{k-1}}, i = 1, 2 \ldots L \\ x_{k-1}^i = \hat{x}_{k-1} - \sqrt{(L+\lambda)P_{k-1}}, i = L+1, L+2, \ldots 2L \end{cases} \tag{18}$$

where L is the length of the state vector, the length of the state vector in this paper is 3 and the weight value calculation is shown in Equation (19):

$$\begin{cases} \lambda = \alpha^2(L+k_i) - L \\ W_m^0 = \frac{\lambda}{L+\lambda}, W_m^i = \frac{1}{2(L+\lambda)}, i = 1, 2 \ldots 2L \\ W_c^0 = \frac{\lambda}{L+\lambda} + 1 - \alpha^2 + \beta, W_c^i = \frac{1}{2(L+\lambda)}, i = 1, 2 \ldots 2L \end{cases} \tag{19}$$

where $\alpha = 0.01, k_i = 0, \beta = 2$.

(3) Time update.

Update predicted status value $\bar{x}_k$:

$$x_{k|k-1}^i = F(x_{k-1}^i) \tag{20}$$

$$\bar{x}_k = \sum_{i=0}^{2L} W_m^i x_k^i \tag{21}$$

Update predicted observation $\bar{y}_k$.

$$y_{k|k-1}^i = G(x_{k|k-1}^i) \tag{22}$$

$$\overline{y}_k = \sum_{i=0}^{2L} W_m^i [G(x_{k|k-1}^i) + v] = \sum_{i=0}^{2L} W_m^i y_{k|k-1}^i \qquad (23)$$

Update system covariance prediction value $P_{xx|k}$.

$$P_{xx|k} = \sum_{i=0}^{2L} \left(W_c^i (x_{k|k-1}^i - \overline{x}_k)(x_{k|k-1}^i - \overline{x}_k)^T \right) + Q_{k-1} \qquad (24)$$

Calculate innovation value d_k and innovation variance value C_{d_k}.

$$d_k = y_k - \overline{y}_k \qquad (25)$$

$$C_{d_k} = \begin{cases} \frac{k-1}{k} C_{d_{k-1}} + \frac{1}{k} d_k d_k^T & k \leq W \\ \frac{1}{W} \sum_{i=k-W+1}^{k} d_i d_i^T & k > W \end{cases} \qquad (26)$$

Update system noise covariance Q_k.

$$Q_k = K_{k-1} C_{d_k} K_{k-1}^T \qquad (27)$$

(4) Status update.

Update observation covariance prediction value $P_{yy|k}$.

$$P_{yy|k} = \sum_{i=0}^{2L} \left(W_c^i (y_{k|k-1}^i - \overline{y}_k)(y_{k|k-1}^i - \overline{y}_k)^T \right) + R_{k-1} \qquad (28)$$

Update covariance $P_{xy|k}$.

$$P_{xy|k} = \sum_{i=0}^{2L} W_c^i (x_{k|k-1}^i - \overline{x}_k)(y_{k|k-1}^i - \overline{y}_k)^T \qquad (29)$$

Calculate Kalman gain K_k.

$$K_k = \frac{P_{xy|k}}{P_{yy|k}} \qquad (30)$$

Update estimated state value $\hat{x}_k$.

$$\hat{x}_k = \overline{x}_k + K_k (y_k - \overline{y}_k) \qquad (31)$$

Update estimated observation $\hat{y}_k$.

$$\hat{y}_k = H_k \hat{x}_k \qquad (32)$$

Update error covariance P_k.

$$P_k = P_{xx|k} - K_k P_{yy|k} K^T \qquad (33)$$

Calculate the residual value r_k and the residual variance value C_{r_k}.

$$r_k = y_k - \hat{y}_k \qquad (34)$$

$$C_{r_k} = \begin{cases} \frac{k-1}{k} C_{d_{r-1}} + \frac{1}{k} r_k r_k^T & k \leq W \\ \frac{1}{W} \sum_{i=k-W+1}^{k} r_i r_i^T & k > W \end{cases} \qquad (35)$$

Update the observation noise covariance R_k.

$$R_k = C_{r_k} + H_k P_k H_k^T \tag{36}$$

The AUKF algorithm flow is shown in Figure 11.

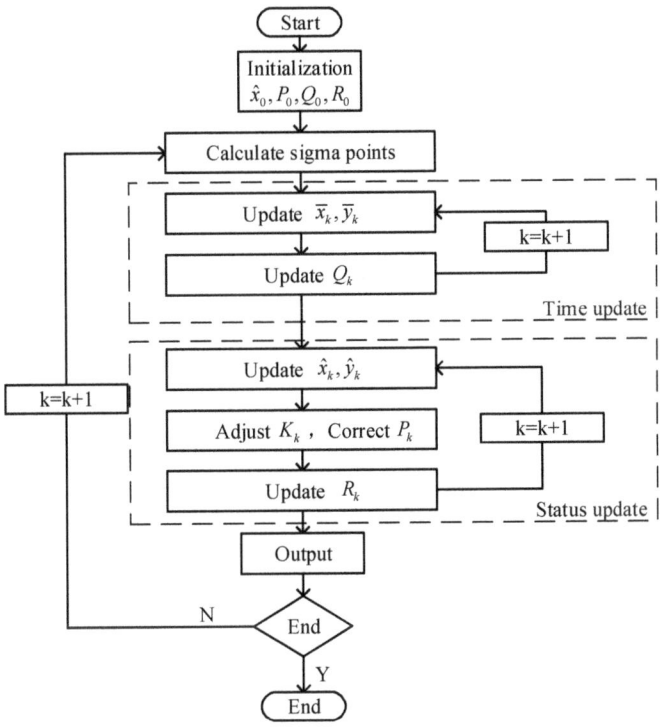

Figure 11. The flow of the AUKF algorithm.

3.2.1. Adaptive System Noise Covariance Q_k

From Equations (18), (21), (24), (27) and (33), it can be seen that when the Q_k value is too large, system covariance prediction $P_{xx|k}$ increases, so that the next predicted state value $\bar{x}_{k+1}$ becomes larger, which eventually leads to the estimated state value $\hat{x}_{k+1}$ being too large, which increases the SOC estimation error. Therefore, the system noise covariance Q_k can be updated in real time to correct the influence of the system error on the estimation result.

The innovation d_k at time k is defined as the difference between the actual observation value y_k and the predicted observation value $\bar{y}_k$. The expression of innovation is shown in Equation (37):

$$d_k = y_k - \bar{y}_k \tag{37}$$

According to the moving window method, the variance of innovation C_{d_k} is calculated as:

$$C_{d_k} = \begin{cases} \frac{k-1}{k} C_{d_{k-1}} + \frac{1}{k} d_k d_k^T & k \leq W \\ \frac{1}{W} \sum_{i=k-W+1}^{k} d_i d_i^T & k > W \end{cases} \tag{38}$$

where W is the length of the moving window. Through the innovation variance C_{d_k}, the system noise covariance Q_k can be calculated [36] as shown in Equation (39),

$$Q_k = K_{k-1} C_{d_{k-1}} K_{k-1}^T \qquad (39)$$

Since the system state variable has a dimension of 3, Q_k is a 3×3 symmetric matrix. This paper will represent Q_k as $Q_k = \begin{bmatrix} Q_{11} & Q_{12} & Q_{13} \\ Q_{21} & Q_{22} & Q_{23} \\ Q_{31} & Q_{32} & Q_{33} \end{bmatrix}$, where $Q_{12} = Q_{21}, Q_{13} = Q_{31}, Q_{23} = Q_{32}$. The Q_k value when using the AUKF algorithm to estimate the SOC in MATLAB is shown in Figure 12.

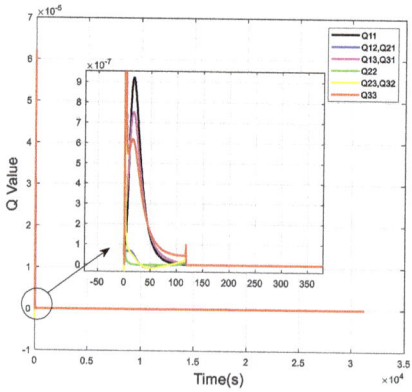

Figure 12. Q value of AUKF algorithm.

It can be seen from Figure 12 that since the initial value of SOC is uncertain, the system error is relatively large at this time, so the system noise covariance Q_k is relatively large. By calculating the value of innovation d_k, and then updating Q_k in real time to correct the error covariance P_k in time, the system noise is corrected in time, and the value of Q_k approaches to zero.

3.2.2. Adaptive Observation Noise Covariance R_k

From Equations (28), (30) and (33), it can be seen that the value of R_k determines the weight of the observation value to the estimated result. When the R_k value increases, the filter gain K_k decreases, resulting in the effect of the observation value on the estimated state value becoming smaller. Conversely, when the value of R_k decreases, the filter gain K_k will increase, which will increase the proportion of the observation value in the estimated state value. Therefore, the observation noise covariance R_k adjusts the Kalman gain K_k in real time to change the proportion of the predicted observation value in the estimation result, thereby reducing the influence of the observation noise on the estimation result.

The residual r_k at time k is defined as the difference between the actual observation value y_k and the estimated observation value $\hat{y}_k$. The expression of the residual is shown in Equation (40):

$$r_k = y_k - \hat{y}_k \qquad (40)$$

According to the moving window method, the variance of residual C_{r_k} is calculated as:

$$C_{r_k} = \begin{cases} \frac{k-1}{k} C_{r_{k-1}} + \frac{1}{k} r_k r_k^T & k \leq W \\ \frac{1}{W} \sum_{i=k-W+1}^{k} r_i r_i^T & k > W \end{cases} \qquad (41)$$

Through the residual variance C_{r_k}, the observation noise covariance R_k can be calculated [37] as shown in Equation (42),

$$R_k = C_{r_k} + H_k P_{k-1} H_k^T \qquad (42)$$

The R_k value when using the AUKF algorithm to estimate the SOC in MATLAB is shown in the Figure 13.

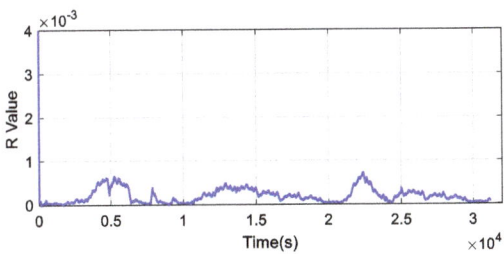

Figure 13. R value of AUKF algorithm.

It can be seen from Figure 13 that the value of R_k fluctuates in a small range. The residual r_k calculates the difference between the actual observation value and the estimated observation value, and then realizes the real-time update of R_k, and then adjusts the Kalman gain K_k to achieve the optimal estimation.

4. Comparison of SOC Estimation Algorithms

The battery SOC was estimated using the unscented Kalman filter algorithm; Q_k and R_k in the adaptive unscented Kalman filter algorithm were analyzed and simulated—see Section 3. In this section, we describe how the AUKF algorithm was used to estimate the battery SOC under different load cycles and different initial SOC values. The results of SOC estimation using AUKF algorithm and the results of SOC estimation using UKF algorithm are compared and analyzed.

4.1. Under Pulsed Discharge Conditions

We carried on the simulation experiment in MATLAB, using the AUKF algorithm to carry on the SOC estimation experiment under the pulsed discharge condition. Firstly, the initial values of the AUKF algorithm were set as follows: $x_0 = [0\ 0\ SOC]^T$, $P_0 = diag([10^{-5}, 10^{-5}, 10^{-3}])$, $W = 1180$, $Q = 10^{-7} \times eye(3)$, $R = 1$. Then, the AUKF algorithm was used to estimate the SOC under pulsed discharge conditions. The robustness of the proposed AUKF algorithm was tested under different initial SOC conditions. Finally, the results of SOC estimation using UKF algorithm and AUKF algorithm were compared and analyzed.

Experiments and analyses were performed under pulsed discharge conditions. For the initial SOC = 0.4, the comparison between the results estimated using the UKF algorithm and the AUKF algorithm is shown in Figure 14. For the initial SOC = 0.6, the comparison between the results estimated using the UKF algorithm and the AUKF algorithm is shown in Figure 15. For the initial SOC = 0.8, the comparison of the results estimated using the UKF algorithm and the AUKF algorithm is shown in Figure 16.

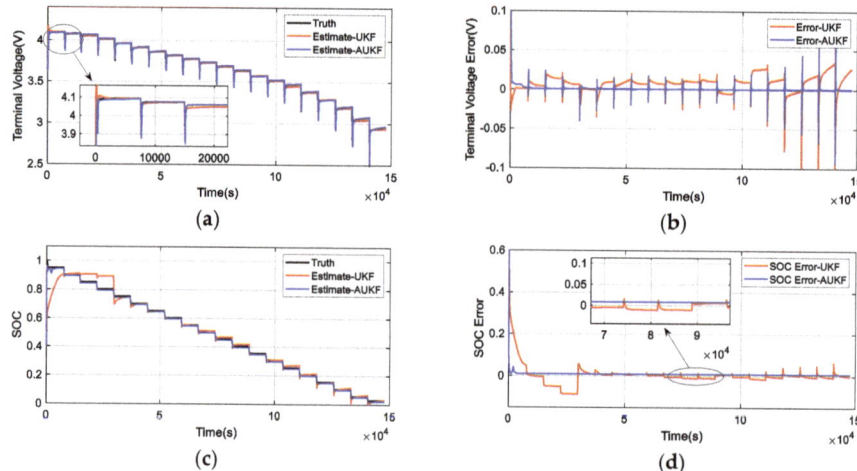

Figure 14. For the initial SOC = 0.4, the comparison of the estimation results of the UKF algorithm and the AUKF algorithm under pulsed discharge conditions. (**a**) Comparison of terminal voltage. (**b**) Comparison of terminal voltage errors. (**c**) Comparison of SOC. (**d**) Comparison of SOC errors.

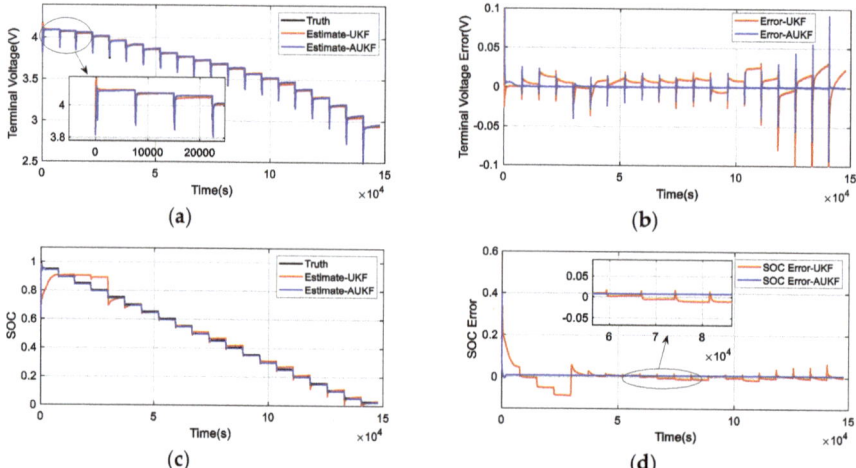

Figure 15. For the initial SOC = 0.6, the comparison of the estimation results of the UKF algorithm and the AUKF algorithm under pulsed discharge conditions. (**a**) Comparison of terminal voltage. (**b**) Comparison of terminal voltage errors. (**c**) Comparison of SOC. (**d**) Comparison of SOC errors.

From Figures 14a, 15a and 16a, it can be seen that under pulsed discharge conditions, the value of the battery terminal voltage estimated by the AUKF algorithm is closer to the true value than the value estimated by the UKF algorithm value. From Figures 14b, 15b and 16b, it can be seen that under pulsed discharge conditions, the terminal voltage error value estimated by the AUKF algorithm is smaller than the terminal voltage error value estimated by the UKF algorithm. Additionally, the error value of the terminal voltage estimated by the AUKF algorithm is relatively small. According to Figures 14c, 15c and 16c, it can be seen that the SOC value estimated using the AUKF algorithm is closer to the true value. From Figures 14d, 15d and 16d, it can be seen that the SOC estimation error of AUKF is smaller than that of UKF, and the convergence speed of AUKF algorithm is faster. In summary, under pulsed discharge conditions and different initial SOC conditions, the robustness of the AUKF algorithm for estimating the SOC of the battery is better than that of the UKF algorithm.

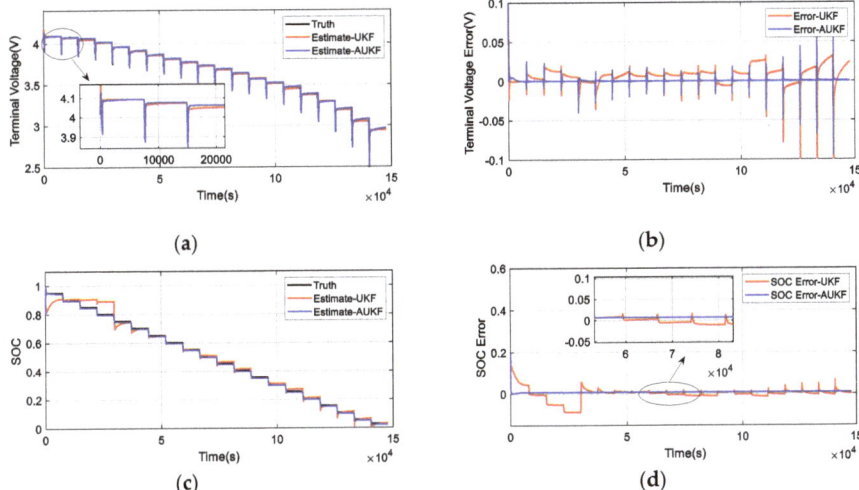

Figure 16. For the initial SOC = 0.8, the comparison of the estimation results of the UKF algorithm and the AUKF algorithm under pulsed discharge conditions. (**a**) Comparison of terminal voltage. (**b**) Comparison of terminal voltage errors. (**c**) Comparison of SOC. (**d**) Comparison of SOC errors.

4.2. Under UDDS Discharge Conditions

We carried on the simulation experiment in MATLAB, using AUKF algorithm to carry on the SOC estimation experiment under the UDDS discharge condition. Firstly, the initial values of the AUKF algorithm were set as follows: $x_0 = [0\ 0\ SOC]^T$, $P_0 = diag([10^{-5}, 10^{-5}, 10^{-3}])$, $W = 1180$, $Q = 10^{-7} \times eye(3)$, $R = 1$. Then, the AUKF algorithm was used to estimate the SOC under UDDS discharge conditions. The robustness of the proposed AUKF algorithm was tested under different initial SOC conditions. Finally, the results of SOC estimation using UKF algorithm and AUKF algorithm were compared and analyzed.

Experiments and analyses were performed under UDDS discharge conditions. For the initial SOC = 0.4, the comparison between the results estimated using the UKF algorithm and the AUKF algorithm is shown in Figure 17. For the initial SOC = 0.6, the comparison between the results estimated using the UKF algorithm and the AUKF algorithm is shown in Figure 18. For the initial SOC = 0.8, the comparison of the results estimated using the UKF algorithm and the AUKF algorithm is shown in Figure 19.

From Figures 17a, 18a and 19a, it can be seen that under UDDS discharge conditions, the value of the battery terminal voltage estimated by the AUKF algorithm is closer to the true value than the value estimated by the UKF algorithm. From Figures 17b, 18b and 19b, it can be seen that under UDDS discharge conditions, the terminal voltage error value estimated by the AUKF algorithm is smaller than the terminal voltage error value estimated by the UKF algorithm. Additionally, the error value of the terminal voltage estimated by the AUKF algorithm is relatively stable. According to Figures 17c, 18c and 19c, it can be concluded that the SOC value estimated using the AUKF algorithm is closer to the true value. From Figures 17d, 18d and 19d, it can be seen that the SOC estimation error of AUKF is smaller than that of UKF, and the convergence speed of AUKF algorithm is faster. In summary, under UDDS discharge conditions and under different initial SOC conditions, the robustness of the AUKF algorithm for estimating the SOC of the battery is better than the UKF algorithm.

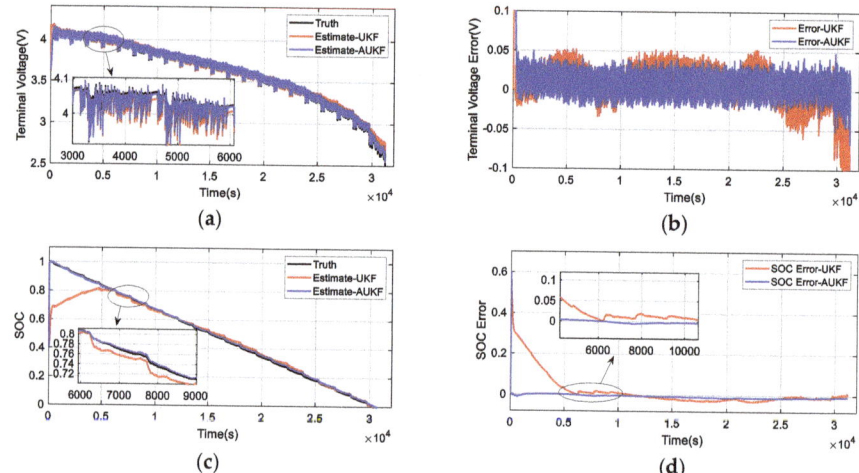

Figure 17. For the initial SOC = 0.4, the comparison of the estimation results of the UKF algorithm and the AUKF algorithm under UDDS discharge conditions. (**a**) Comparison of terminal voltage. (**b**) Comparison of terminal voltage errors. (**c**) Comparison of SOC. (**d**) Comparison of SOC errors.

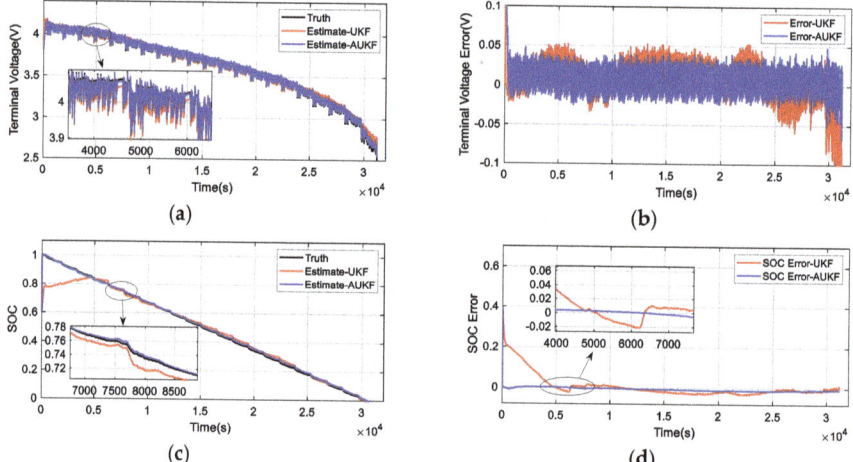

Figure 18. For the initial SOC = 0.6, the comparison of the estimation results of the UKF algorithm and the AUKF algorithm under UDDS discharge conditions. (**a**) Comparison of terminal voltage. (**b**) Comparison of terminal voltage errors. (**c**) Comparison of SOC. (**d**) Comparison of SOC errors.

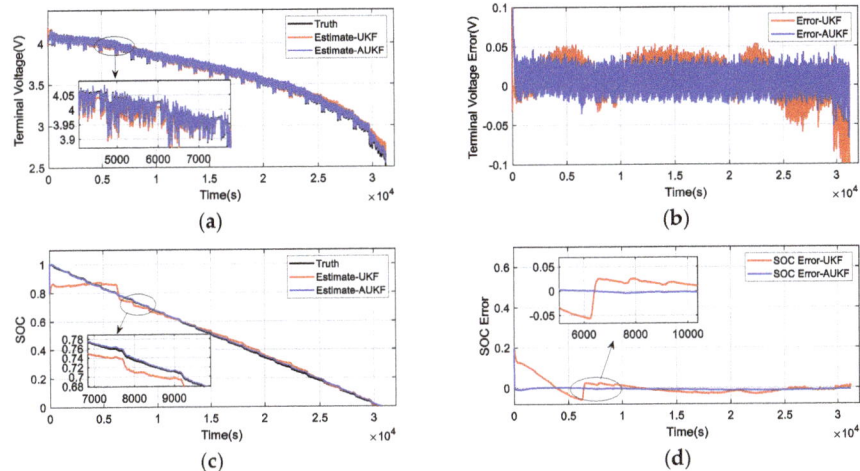

Figure 19. For the initial SOC = 0.8, the comparison of the estimation results of the UKF algorithm and the AUKF algorithm under UDDS discharge conditions. (**a**) Comparison of terminal voltage. (**b**) Comparison of terminal voltage errors. (**c**) Comparison of SOC. (**d**) Comparison of SOC errors.

In order to further verify the accuracy of the SOC estimation by the AUKF algorithm, the terminal voltage value and the SOC value estimated by the UKF algorithm and the AUKF algorithm were subjected to error analysis. The error results of the battery SOC estimation under different load cycles and different initial SOC values were averaged. The error analysis results of the terminal voltage are shown in Table 5. The error analysis results of the estimated SOC are shown in Table 6.

Table 5. Terminal voltage error.

Algorithm	Error Type	MAE	RMSE
UKF		1.33%	1.95%
AUKF		0.54%	0.9%

Table 6. SOC error.

Algorithm	Error Type	MAE	RMSE
UKF		2.9%	3.3%
AUKF		0.63%	0.7%

The mean absolute error (MAE) can avoid the problem of the deviations cancelling each other out, and can well describe the degree of data dispersion. The root mean square error (RMSE) measures the deviation between the observation value and the true value, and can well reflect the accuracy of the measurement. Table 5 shows that the MAE of the terminal voltage estimated by AUKF was smaller, indicating that the terminal voltage estimated by AUKF is less discrete than that of UKF algorithm; the RMSE of terminal voltage estimated by AUKF algorithm was smaller than that of UKF algorithm. Table 6 shows that the SOC value estimated by the AUKF algorithm had a smaller MAE than that of UKF algorithm. The RMSE of SOC estimated by AUKF algorithm was 2.6% smaller than that of UKF algorithm. The above error analysis results indicate that the accuracy of SOC estimation using AUKF algorithm is better than that of UKF algorithm.

5. Conclusions

In this paper, an adaptive unscented Kalman filter algorithm was designed to estimate the SOC of a lithium cobalt oxide battery. The second-order RC equivalent circuit model was used for nonlinear modeling of batteries. The least square method was used to identify the parameters of the battery model and for simulations in MALAB according to the battery voltage characteristics. The established model was verified under pulsed discharge conditions; the model error is 0.8%, which provides an accurate model for SOC estimation using AUKF algorithm. The system noise covariance Q value and observation noise covariance R value in the unscented Kalman filter algorithm were analyzed in this paper. The AUKF algorithm updates the system noise covariance Q and the observation noise covariance R in real time by monitoring the changes of the innovation and residual in the filter to adjust the filter gain and achieve the optimal estimate. The AUKF algorithm and UKF algorithm were used for SOC estimation under different load cycles and different initial SOC values. The estimated result of AUKF algorithm was more accurate, and the convergence rate of filtering was faster than that of UKF algorithm. To further verify the effectiveness of the AUKF algorithm, the estimated error of the terminal voltage and the estimated error of the SOC were analyzed. The error results show that the error of SOC estimation using AUKF algorithm was 0.7%, which was 2.6% smaller than that of SOC estimation using UKF algorithm.

Author Contributions: J.L. and X.W. proposed the innovative idea; J.L. conceived the algorithm and wrote the first draft; Y.L. improved the algorithm; J.L. performed the experiments; J.L. and X.W. analyzed the results; Y.F. and B.J. provided writing advice; J.L. approved the final manuscript. All authors have read and agreed to the published version of the manuscript.

Funding: The research received no external funding.

Acknowledgments: We gratefully acknowledge the technical assistance of DL850E ScopeCorder.

Conflicts of Interest: The authors declare no conflict of interest.

References

1. Gruosso, G.; Gajani, G.S.; Ruiz, F.; Valladolid, J.D.; Patino, D. A virtual sensor for electric vehicles' state of charge estimation. *Electronics* **2020**, *9*, 278. [CrossRef]
2. Yuan, W.; Jeong, S.; Sean, W.; Chiang, Y. Development of Enhancing Battery Management for Reusing Automotive Lithium-Ion Battery. *Energies* **2020**, *13*, 3306. [CrossRef]
3. Yang, Y.; Tan, Z.; Ren, Y. Research on factors that influence the fast charging behavior of private battery electric vehicles. *Sustainability* **2020**, *12*, 3539. [CrossRef]
4. Feng, F.; Teng, S.; Liu, K.; Xie, J.; Xie, Y.; Liu, B.; Li, K. Co-estimation of lithium-ion battery state of charge and state of temperature based on a hybrid electrochemical-thermal-neural-network model. *J. Power Sources* **2020**, *455*, 227935. [CrossRef]
5. Li, Y.; Chattopadhyay, P.; Xiong, S.; Ray, A.; Rahn, C.D. Dynamic data-driven and model-based recursive analysis for estimation of battery state-of-charge. *Appl. Energy* **2016**, *184*, 266–275. [CrossRef]
6. Chang, L.; Wang, C.; Zhang, C.; Xiao, L.; Cui, N.; Li, H.; Qiu, J. A novel fast capacity estimation method based on current curves of parallel-connected cells for retired lithium-ion batteries in second-use applications. *J. Power Sources* **2020**, *459*, 227901. [CrossRef]
7. Vo, T.T.; Chen, X.; Shen, W.; Kapoor, A. New charging strategy for lithium-ion batteries based on the integration of Taguchi method and state of charge estimation. *J. Power Sources* **2015**, *273*, 413–422. [CrossRef]
8. Chen, X.; Lei, H.; Xiong, R.; Shen, W.; Yang, R. A novel approach to reconstruct open circuit voltage for state of charge estimation of lithium ion batteries in electric vehicles. *Appl. Energy* **2019**, *255*, 113758. [CrossRef]
9. Stolze, C.; Hager, M.D.; Schubert, U.S. State-of-charge monitoring for redox flow batteries: A symmetric open-circuit cell approach. *J. Power Sources* **2019**, *423*, 60–67. [CrossRef]
10. Yang, F.; Li, W.; Li, C.; Miao, Q. State-of-charge estimation of lithium-ion batteries based on gated recurrent neural network. *Energy* **2019**, *175*, 66–75. [CrossRef]
11. Bian, C.; He, H.; Yang, S. Stacked bidirectional long short-term memory networks for state-of-charge estimation of lithium-ion batteries. *Energy* **2020**, *191*, 116538. [CrossRef]

12. Martino, L.; Read, J.; Elvira, V.; Louzada, F. Cooperative parallel particle filters for online model selection and applications to urban mobility. *Digit. Signal. Process.* **2017**, *60*, 172–185. [CrossRef]
13. Martino, L.; Elvira, V.; Camps-Valls, G. Group Importance Sampling for particle filtering and MCMC. *Digit. Signal. Process.* **2018**, *82*, 133–151. [CrossRef]
14. Mawonou, K.S.R.; Eddahech, A.; Dumur, D.; Beauvois, D.; Godoy, E. Improved state of charge estimation for Li-ion batteries using fractional order extended Kalman filter. *J. Power Sources* **2019**, *435*, 226710. [CrossRef]
15. Linghu, J.; Kang, L.; Liu, M.; Luo, X.; Feng, Y.; Lu, C. Estimation for state-of-charge of lithium-ion battery based on an adaptive high-degree cubature Kalman filter. *Energy* **2019**, *189*, 116204. [CrossRef]
16. Guo, F.; Hu, G.; Xiang, S.; Zhou, P.; Hong, R.; Xiong, N. A multi-scale parameter adaptive method for state of charge and parameter estimation of lithium-ion batteries using dual Kalman filters. *Energy* **2019**, *178*, 79–88. [CrossRef]
17. Zheng, Y.; Gao, W.; Ouyang, M.; Lu, L.; Zhou, L.; Han, X. State-of-charge inconsistency estimation of lithium-ion battery pack using mean-difference model and extended Kalman filter. *J. Power Sources* **2018**, *383*, 50–58. [CrossRef]
18. Tian, Y.; Lai, R.; Li, X.; Xiang, L.; Tian, J. A combined method for state-of-charge estimation for lithium-ion batteries using a long short-term memory network and an adaptive cubature Kalman filter. *Appl. Energy* **2020**, *265*, 114789. [CrossRef]
19. Partovibakhsh, M.; Liu, G. An adaptive unscented kalman filtering approach for online estimation of model parameters and state-of-charge of lithium-ion batteries for autonomous mobile robots. *IEEE Trans. Control. Syst. Technol.* **2015**, *23*, 357–363. [CrossRef]
20. Wang, S.; Fernandez, C.; Yu, C.; Fan, Y.; Cao, W.; Stroe, D.I. A novel charged state prediction method of the lithium ion battery packs based on the composite equivalent modeling and improved splice Kalman filtering algorithm. *J. Power Sources* **2020**, *471*, 228450. [CrossRef]
21. Pérez, G.; Garmendia, M.; Reynaud, J.F.; Crego, J.; Viscarret, U. Enhanced closed loop State of Charge estimator for lithium-ion batteries based on Extended Kalman Filter. *Appl. Energy* **2015**, *155*, 834–845. [CrossRef]
22. Smiley, A.; Plett, G.L. An adaptive physics-based reduced-order model of an aged lithium-ion cell, selected using an interacting multiple-model Kalman filter. *J. Energy Storage* **2018**, *19*, 120–134. [CrossRef]
23. Plett, G.L. Extended Kalman filtering for battery management systems of LiPB-based HEV battery packs—Part 2. Modeling and identification. *J. Power Sources* **2004**, *134*, 262–276. [CrossRef]
24. Plett, G.L. Extended Kalman filtering for battery management systems of LiPB-based HEV battery packs—Part 3. State and parameter estimation. *J. Power Sources* **2004**, *134*, 277–292. [CrossRef]
25. He, H.; Xiong, R.; Peng, J. Real-time estimation of battery state-of-charge with unscented Kalman filter and RTOS μCOS-II platform. *Appl. Energy* **2016**, *162*, 1410–1418. [CrossRef]
26. Sun, F.; Hu, X.; Zou, Y.; Li, S. Adaptive unscented Kalman filtering for state of charge estimation of a lithium-ion battery for electric vehicles. *Energy* **2011**, *36*, 3531–3540. [CrossRef]
27. Gustafsson, F.; Hendeby, G. Some relations between extended and unscented Kalman filters. *IEEE Trans. Signal. Process.* **2012**, *60*, 545–555. [CrossRef]
28. Chen, K.; Yu, J. Short-term wind speed prediction using an unscented Kalman filter based state-space support vector regression approach. *Appl. Energy* **2014**, *113*, 690–705. [CrossRef]
29. Lin, C.; Tang, A.; Xing, J. Evaluation of electrochemical models based battery state-of-charge estimation approaches for electric vehicles. *Appl. Energy* **2017**, *207*, 394–404. [CrossRef]
30. Ringbeck, F.; Garbade, M.; Sauer, D.U. Uncertainty-aware state estimation for electrochemical model-based fast charging control of lithium-ion batteries. *J. Power Sources* **2020**, *470*, 228221. [CrossRef]
31. Seaman, A.; Dao, T.S.; McPhee, J. A survey of mathematics-based equivalent-circuit and electrochemical battery models for hybrid and electric vehicle simulation. *J. Power Sources* **2014**, *256*, 410–423. [CrossRef]
32. Zhang, X.; Lu, J.; Yuan, S.; Yang, J.; Zhou, X. A novel method for identification of lithium-ion battery equivalent circuit model parameters considering electrochemical properties. *J. Power Sources* **2017**, *345*, 21–29. [CrossRef]
33. Bhattacharjee, A.; Roy, A.; Banerjee, N.; Patra, S.; Saha, H. Precision dynamic equivalent circuit model of a Vanadium Redox Flow Battery and determination of circuit parameters for its optimal performance in renewable energy applications. *J. Power Sources* **2018**, *396*, 506–518. [CrossRef]

34. Zhu, Q.; Xiong, N.; Yang, M.L.; Huang, R.S.; Di Hu, G. State of charge estimation for lithium-ion battery based on nonlinear observer: An H∞ method. *Energies* **2017**, *10*, 679. [CrossRef]
35. Piller, S.; Perrin, M.; Jossen, A. Methods for state-of-charge determination and their applications. *J. Power Sources* **2001**, *96*, 113–120. [CrossRef]
36. Zhang, Q.; Yang, Y.; Xiang, Q.; He, Q.; Zhou, Z.; Yao, Y. Noise Adaptive Kalman Filter for Joint Polarization Tracking and Channel Equalization Using Cascaded Covariance Matching. *IEEE Photonics J.* **2018**, *10*, 1–11. [CrossRef]
37. Song, M.; Astroza, R.; Ebrahimian, H.; Moaveni, B.; Papadimitriou, C. Adaptive Kalman filters for nonlinear finite element model updating. *Mech. Syst. Signal. Process.* **2020**, *143*, 106837. [CrossRef]

© 2020 by the authors. Licensee MDPI, Basel, Switzerland. This article is an open access article distributed under the terms and conditions of the Creative Commons Attribution (CC BY) license (http://creativecommons.org/licenses/by/4.0/).

Article

Ultrasonic Health Monitoring of Lithium-Ion Batteries

Yi Wu [1,2,*], Youren Wang [1], Winco K. C. Yung [3] and Michael Pecht [2]

1. College of Automation Engineering, Nanjing University of Aeronautics and Astronautics, Nanjing 211106, China
2. Center for Advanced Life Cycle Engineering (CALCE), University of Maryland, College Park, MD 20742, USA
3. Department of Industrial and System Engineering, The Hong Kong Polytechnic University, Hong Kong 999077, China
* Correspondence: janeyi105@126.com; Tel.: +86-139-1596-3856

Received: 10 April 2019; Accepted: 1 July 2019; Published: 3 July 2019

Abstract: Because of the complex physiochemical nature of the lithium-ion battery, it is difficult to identify the internal changes that lead to battery degradation and failure. This study develops an ultrasonic sensing technique for monitoring the commercial lithium-ion pouch cells and demonstrates this technique through experimental studies. Data fusion analysis is implemented using the ultrasonic sensing data to construct a new battery health indicator, thus extending the capabilities of traditional battery management systems. The combination of the ultrasonic sensing and data fusion approach is validated and shown to be effective for degradation assessment as well as early failure indication.

Keywords: lithium-ion battery; ultrasonic sensing; health monitoring; state of health; failure indication; data fusion

1. Introduction

Lithium-ion batteries are widely used as the power supply for products ranging from portable consumer products to transportation power sources. However, they are prone to experience gradual degradation (capacity fade and resistance increase) or catastrophic failures during usage. The gradual degradation is driven by complex electrochemical side reactions (e.g., active material dissolution, electrode particle cracking, and deterioration of electrode adhesion) over the long period of normal cycling [1]. The catastrophic failures of a battery include a sudden drop in battery capacity, a sudden increase in battery temperature, swelling due to gas generation, and even fire/explosion [2]. A battery can fail catastrophically due to manufacturing defects, mechanical abuse (e.g., shock, or puncture), electrical abuse (e.g., overcharge, or over-discharge), or thermal abuse (e.g., external heating) [3].

Battery degradation and failures result in poor operational availability and, in some cases, safety issues [4–7]. Monitoring battery performance and health is a means to improve the safety and reliability of battery-powered devices. Conventional battery management systems (BMSs) evaluate the health state of a battery by tracking the state of health (SOH, typically defined as the ratio of the maximum deliverable capacity to the initial capacity) using complex state estimation algorithms [8]. However, early detection of battery failure is somewhat difficult based on these SOH estimation techniques because the SOH changes over long periods of time, which makes estimation conservative and slow to react to sudden failures. Furthermore, the conventional BMSs are limited to monitoring the extrinsic parameters (e.g., current, voltage, and temperature) and do not provide insight into the changes inside the battery [9–13].

Due to the complex physiochemical nature of batteries, it is difficult to employ sensors that directly probe the internal chemical properties of the battery (e.g., lithium concentration and electrode properties).

Nevertheless, several state-of-the-art sensors have been developed to monitor the internal physical states of batteries. Fiber optic sensors have been used for internal temperature [14] and strain [15,16] detection. An integrated microsensor has been developed using a micro-electro-mechanical system, which can in situ monitor internal temperature in a coin cell [17]. Although these cutting-edge techniques are promising, they are too costly to be implemented and the effect of these built-in sensors on battery reliability and safety has not been assessed.

Ultrasonic inspection is one of the most widely used approaches in the context of structural health monitoring [18] and nondestructive evaluation [19]. Ultrasonic waves are propagated through the test object, allowing the material properties, internal damage, and structural integrity to be accurately monitored in real time. Hence, ultrasonic inspection can potentially be used to probe the underlying material properties changes inside batteries, and thus provide an early indication of failures.

Studies on ultrasonic sensing of batteries, including those by the Ultran Group [20], Sood et al. [21], and Li et al. [22] have reported detection of gas voids, electrolyte nonuniformity, and cracks within the electrode layer of batteries. Ladpli et al. [23–25] and Hsieh et al. [26] studied ultrasonic trends associated with the change of the state of charge (SOC, the charge remaining in the battery with respect to the capacity) and the SOH over charge/discharge cycles. Gold et al. [27] developed a linear model between the ultrasonic transmission signal and the SOC for the SOC determination over one cycle. Similarly, Davies et al. [28] developed a support vector regression (SVR) model to predict the battery SOC and SOH using a combination of ultrasonic and voltage data. These studies focused on battery performance evaluation and did not consider the sudden failure scenario during operation. However, early prediction of battery sudden failures is crucial for improving battery safety because sudden failures often induce catastrophic events, such as fire or explosion.

This study enhances the state of the art by expanding the use of ultrasonic sensing for early indication of battery sudden failure. Two types of battery tests were conducted to investigate the evolution of the ultrasonic waves—a cycling test and an abusive test (overcharge test). During the cycling test, batteries were cycled under different conditions, and correlations between the ultrasonic features and the health state of the battery were identified using the Spearman correlation. During the abusive test, the ultrasonic signal was continuously monitored, and the evolution of the ultrasonic features within the overcharge process was analyzed. Then, a data fusion approach and a health indicator (HI) were developed to quantify the battery health state and indicate catastrophic failure induced by overcharge. Section 2 describes the theoretical background for battery ultrasonic sensing, Section 3 presents the battery cycling and abusive tests, and Section 4 discusses the ultrasonic test results. Section 5 describes the developed battery health monitoring method with one case study. Conclusions are presented in Section 6.

2. Ultrasonic Sensing for Lithium-Ion Batteries

There are two modes for the ultrasonic inspection. The first is the pulse-echo mode, where the ultrasonic signal is sent and received by the same transducer, and the second is the through-transmission mode, where the ultrasonic signal transmits through the object and is received by the second transducer [21]. The through-transmission mode needs two transducers that are placed on opposite sides of the battery, which requires more access to the battery than the pulse-echo mode and increases the cost. Thus, the ultrasonic testing of the batteries in this study was conducted in the pulse-echo mode as shown in Figure 1a.

To conduct the ultrasonic test, the transducer sends a compressional pulse through the battery. The signal propagates through the battery, which consists of many layers stacked together, e.g., positive electrode/separator/negative electrode/separator [29]. The ultrasonic velocity in each layer depends on the material properties. For an isotropic elastic material, the ultrasonic velocity is:

$$V = \sqrt{\frac{\frac{4}{3}G + K}{\rho}} \qquad (1)$$

where G is the material shear modulus, K is the bulk modulus, and ρ is the material density.

At the interface between two different layers, part of the ultrasonic signal is transmitted through the interface, while the rest is reflected. The greater the ultrasonic impedance (Z) mismatch between two layers, the greater the percentage of energy reflected at the interface. The ultrasonic impedance (Z) of a material is defined as the product of its density (ρ) and ultrasonic velocity (V).

$$Z = \rho \times V \tag{2}$$

Figure 1b shows a typical ultrasonic response signal for a commercial lithium-ion battery. The ultrasonic signal travels through the battery. Part of the signal reflects different interfaces inside the battery, which forms the oscillation in the first 6 µs, while the other part of the signal reaches the bottom of the battery and then reflects back to the top, which is highlighted as the first echo signal. Note that the reflected signal causes a secondary reflection when reaching the top, and thus forms repeated echoes, as labeled in Figure 1b. These echoes have similar shapes and time intervals. The amplitude of the echo peak decreases as the travel time increases because the ultrasonic signal attenuates when it propagates through the layers inside the battery due to reflections and viscoelastic losses.

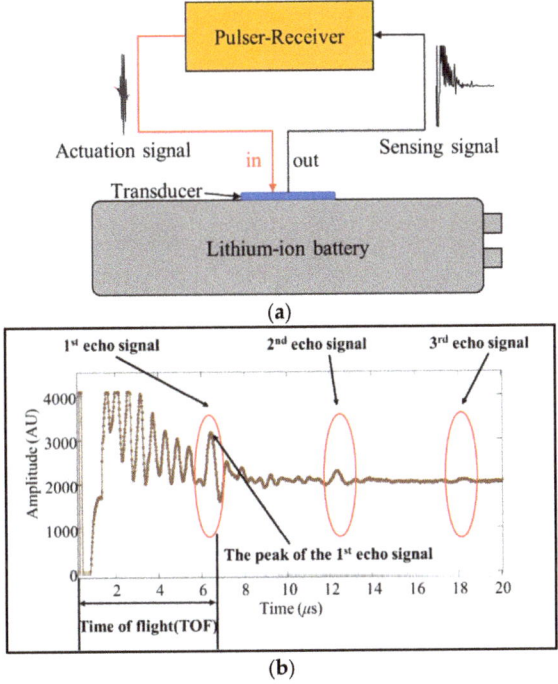

Figure 1. (a) Schematic for battery ultrasonic sensing and (b) representative ultrasonic response signal of a lithium-ion battery.

Two features of interest, i.e., the maximum amplitude of the first echo (defined as PA) and the time-of-flight (TOF), are extracted from the ultrasonic response signal and used to analyze the performance of the battery. The essence of battery degradation is the change of battery material properties (density and modulus), which leads to the change of both ultrasonic impedance and ultrasonic velocity. To indicate the change of ultrasonic impedance, the amplitude of the reflected ultrasonic signal is used because they are closely related. Specifically, we chose the maximum amplitude of the first echo (PA) to represent the ultrasonic amplitude due to its uniqueness and representativeness. Regarding the change of ultrasonic velocity, we chose the TOF as the ultrasonic velocity indicator. The TOF (time between the actuation signal and the PA) reflects the travel time of the ultrasonic signal inside the battery, and thus represents the change of ultrasonic velocity and/or thickness.

3. Experimental Setup

Experiments were performed on two types of commercial $LiCoO_2$/graphite pouch batteries to study the evolution of ultrasonic signals. The first battery type (type A) had a rated capacity of 1.8 Ah, and the second type (type B) had a rated capacity of 0.7 Ah. The voltage range for both types was the same, from 2.75 V to 4.2 V. Two samples were tested for each type.

In the experiment (see Figure 2), a thermocouple and a piezoelectric transducer (PZT-5A, 1 MHz) were mounted on the surface of the batteries with epoxy, instead of the glycerin. This provided clean signals and reduced signal variability. No extra pressure was applied to the batteries. The charge/discharge test was conducted using an Arbin BT2000 battery tester, and the current and voltage of the batteries were recorded. The battery surface temperature was recorded with an Agilent 34970A (Santa Clara, CA, USA) data logger. A Lecoeur ultrasonic device (pulser-receiver) was used to send and receive the ultrasonic signal. The piezoelectric transducer was actuated with a 50 ns pulse. The amplitude of the actuation signal was 100 V. The sampling frequency for the received signal was 80 MHz.

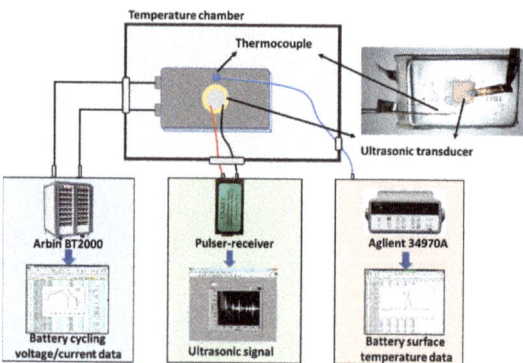

Figure 2. Battery test bench.

To investigate the feasibility of ultrasonic sensing, several cycling tests and an overcharge test were designed and conducted at 45 °C in a temperature chamber to induce the accelerated aging and failure of the batteries. A test temperature of 45 °C was chosen because it is higher than room temperature but does not cause catastrophic failure during normal cycling conditions based on our previous experience. The purpose of these tests was to find the relationship between the battery degradation and ultrasonic signal in a short testing time. The test procedures are summarized in Table 1.

Table 1. Battery test methods.

Test Type	Test Method	Test Description
Cycling test	Normal cycling	CC charge at 0.5C until 4.2 V, then CV charge until current <0.05C Rest 10 min Discharge at 1C to 2.75 V Rest 10 min
	Overcharge cycling	CC charge at 0.5C until 4.5 V, then CV charge until current <0.05C Rest 10 min Discharge at 1C to 2.75 V Rest 10 min
Abusive test	Overcharge	CC charge at 0.5C until 5 V CV charge until the battery swells

For the cycling test, all the batteries were discharged using 1C (C/x is a current rate at which the battery will be fully charged in x hours) to 2.75 V, whereas, different charge profiles were applied to different types of batteries. Type A batteries (A1/A2) were charged using normal charge profiles, i.e., they were charged using 0.5C constant current (CC) to 4.2 V and kept at 4.2 V constant voltage (CV) until the current dropped to 0.05C. Type B batteries (B1/B2) were charged using overcharge profiles, i.e., they were charged using 0.5C to 4.5 V and kept at 4.5 V until the current dropped to 0.05C. The ultrasonic signal was measured at 0% SOC after every cycle.

After a period of cycling test, a forced overcharge test was conducted on battery A1 to understand the change in the ultrasonic signal when it came across a sudden abuse. Battery A1 was charged using 0.5C to 5 V, then maintained at 5 V until the battery swelled. The ultrasonic measurement was taken every 30 s during testing. In addition, the X-ray images for battery A1 before and after overcharge were measured to provide visualizations for the structural changes inside the battery.

4. Ultrasonic Results for Lithium-Ion Batteries

The ultrasonic results for both the cycling tests and abusive test are presented in this section.

4.1. Ultrasonic Results for the Cycling Tests

Since the SOH is widely used as a way to represent the degree of degradation, the relationships between the SOH and the ultrasonic features were investigated for the cycling tests. Battery A1 was cycled under normal charge/discharge conditions. Figure 3a shows the SOH of battery A1 over the cycle number. The initial capacity of battery A1 was 1.88 Ah. The average capacity fading rate was about 0.02% per cycle. After 100 cycles and 210 cycles, the remaining capacity was 97.69% and 96.02% of the initial capacity, respectively. Because of the regeneration phenomena of lithium-ion batteries during the rest time [30], battery A1 recovered some capacity when the ultrasonic measurement was conducted. The evolution of the ultrasonic signal over the cycle number, in Figure 3b, shows that the waveform of the signal shifts over cycles. The ultrasonic signal at cycle 100 deviates from the signal at cycle 1, which was considered as the baseline signal. The deviation at cycle 210 is more prominent as compared with that at cycle 100.

To further analyze the ultrasonic features (i.e., TOF and PA), they are extracted from each cycle, respectively. Figure 3c shows that the TOF basically increases as the battery degrades. Increasing of TOF indicates decreasing of the ultrasound velocity, which is a result of the changes in the electrode densities and modulus (i.e., the lithium content changes in each electrode) as a battery degrades [28,31]. Note that an increase of TOF could also indicate an increase in the thickness. However, changes of thickness are ignored in this case as they are not the main contributor to TOF changes according to [28]. In addition, the signal amplitude, in Figure 3c, generally increases before cycle 50, followed by a fluctuation after cycle 50. A sharp decrease of signal amplitude occurs at cycle 136, which indicates possible gas generation inside the battery [21].

To determine the capability of the signal to indicate the battery SOH, the Spearman correlation coefficient was employed to calculate the correlation between the ultrasonic features (TOF and PA) and the battery SOH [32]. The Spearman correlation coefficient, rs, is computed as follows:

$$rs = \frac{cov(x_i, y_i)}{\sigma x_i, \sigma y_i} \qquad (3)$$

where $cov(x_i, y_i)$ is the covariance of the rank variables x_i and y_i for the original data x_i and y_i, and σx_i, σy_i are the standard deviations of the rank variables. The rs ranges from -1 to $+1$ and $rs = -1$ ($+1$) indicates that one of the variables is monotonically decreasing (increasing) with the other.

Figure 3d shows the Spearman correlation coefficients between the ultrasonic features and the SOH from the initial cycle to different cycles. The negative correlation coefficient values between the SOH and the TOF show that the TOF increases with the decreasing SOH. In addition, the absolute values of the correlation coefficients are always larger than 0.94, which indicates that the TOF is highly correlated with the SOH over cycles. However, as shown in Figure 3d, a strong correlation (<−0.75) between the PA and the SOH is only seen before cycle 100 but not after that.

Figure 3e shows the SOH versus the TOF for battery A1. A linear model is established by fitting the experimental data. A "closer-to-1" adjusted R-square (R^2) and a "closer-to-0" root mean square error (RMSE) show a better goodness-of-fit of the model. In this case, the R^2 of 0.949 and the RMSE of 0.002677 indicate that there is a linear relationship between the SOH and the TOF for battery A1.

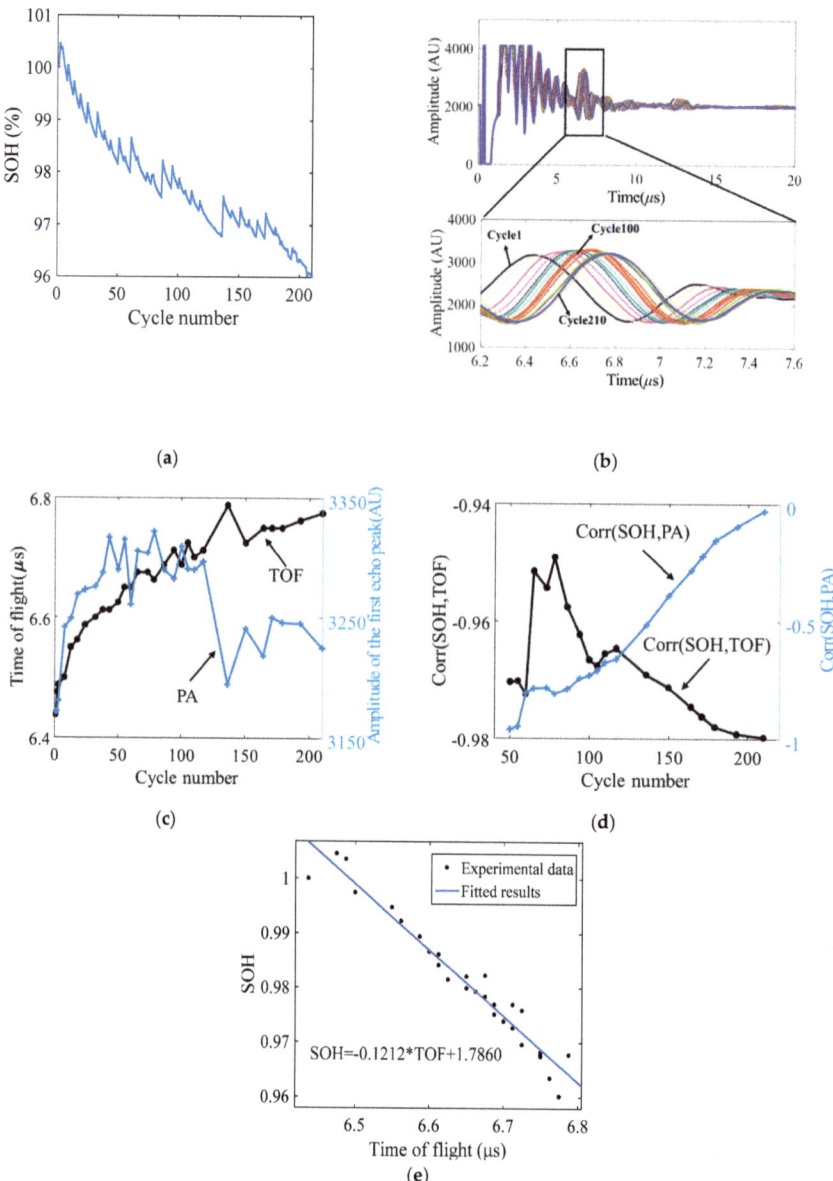

Figure 3. Cycling test results for battery A1: (**a**) cycling performance, (**b**) ultrasonic signals, (**c**) ultrasonic features, (**d**) Spearman correlations between the ultrasonic features and state of health (SOH), and (**e**) plot of SOH as a function of time-of-flight (TOF).

Battery A2 underwent the same cycling test as A1. Figure 4a shows the cycling performance for battery A2. The capacity at cycle 63 is 98.35% of the initial capacity. Figure 4b,c shows the evolution of the TOF and PA, respectively, and their corresponding correlations with the battery SOH were also calculated. Both of these features show high correlations with the SOH. Figure 4d exhibits the relationship between the TOF and SOH of this battery. We observe that the TOF-SOH curve can also be well fitted using a linear model as battery A1.

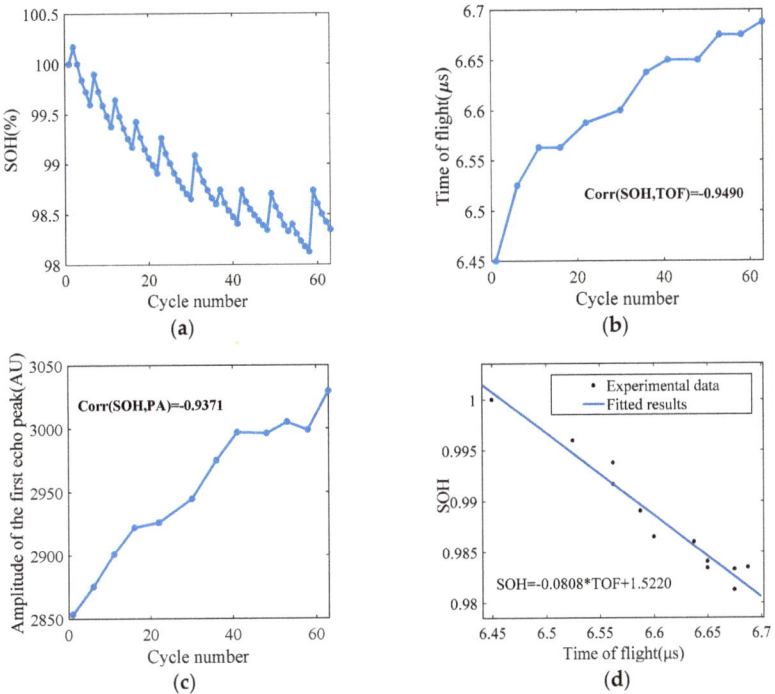

Figure 4. Cycling test results for battery A2: (**a**) cycling performance, (**b**) time-of-flight, (**c**) amplitude of the first echo peak, and (**d**) plot of SOH as a function of TOF.

Batteries B1 and B2 were cycled under overcharge conditions to accelerate battery degradation with more severe physical changes shown than batteries A1 and A2. Figure 5a shows the average capacity fading rates for batteries B1 and B2 were 0.48% and 0.52%, respectively, which are higher as compared with batteries A1 and A2 under normal cycling conditions. Similar to the analysis of batteries A1 and A2, Figure 5b shows the evolution of the TOF for batteries B1 and B2 and the correlation between the TOF and the SOH for each battery is calculated. Figure 5c plots the change in the PA with the cycle number for both these batteries and shows that both these features are highly correlated with the SOH. However, the TOF shows a negative correlation with the SOH, while the PA shows a positive correlation.

According to the ultrasonic analysis from the cycling tests, the evolution of the PA does not appear to be a reliable indicator to characterize the battery SOH because it does not indicate the SOH properly for battery A1 after 100 cycles. In addition, the PA shows a negative correlation with SOH for batteries A1 and A2 and a positive correlation for batteries B1 and B2. Such inconsistent behavior of the PA under different cycling conditions is likely caused by the complex interplay of multiple layers inside the battery because each interface offers the possibility for the ultrasonic signal to split [26]. However, the TOF always shows a strong and consistent correlation with the battery SOH in both the normal and overcharge cycling tests. Therefore, the TOF is a more reliable indicator of battery health than the PA.

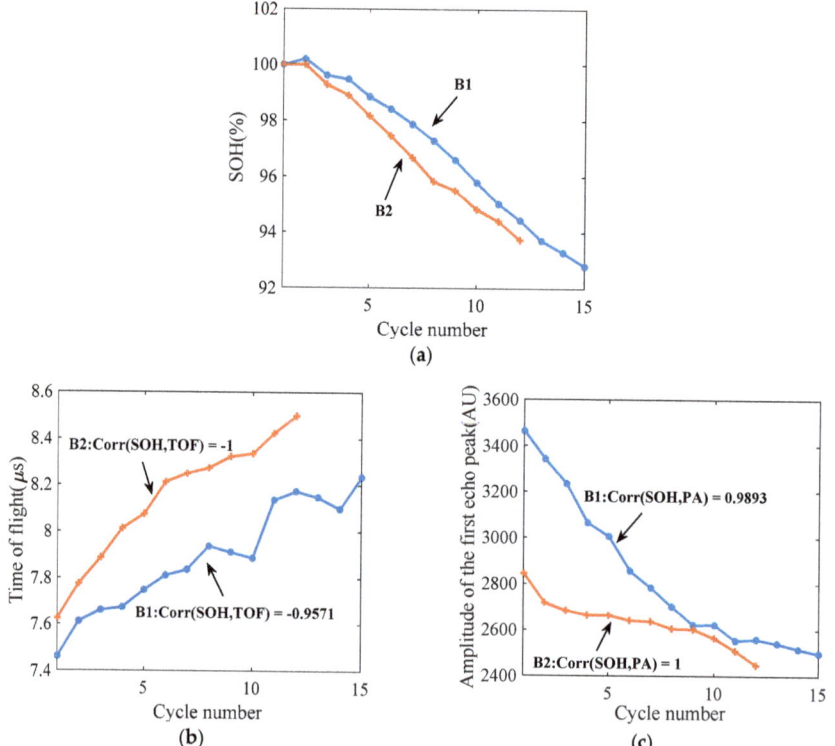

Figure 5. Cycling test results for batteries B1 and B2: (**a**) cycling performance, (**b**) time-of-flight, (**c**) amplitude of the first echo peak.

To further explain the change in the TOF with the battery SOH, Figure 6 shows the evolution of the TOF during a single cycle. As we can see, the TOF decreased during the charging process, while the TOF increased during the discharging process. Referring to [28], as the battery discharges, lithium de-intercalates from the graphite and intercalates into $LiCoO_2$, which decreases the elastic moduli for both graphite and $LiCoO_2$. According to Equation (1), this will decrease the ultrasonic velocity and cause the increase of TOF during the discharging process.

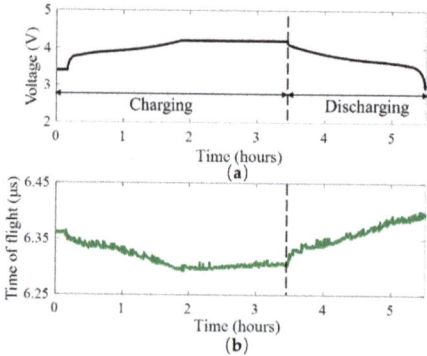

Figure 6. (**a**) Voltage data during a C/2 charging and discharging cycle, (**b**) corresponding evolution of TOF taken every 20 s.

Similarly, active lithium is consumed by side reactions (such as deposition of metallic lithium) when the battery ages. Therefore, at 0% cell SOC, the degree of the cathode lithiation (cathode SOC) may decrease with battery degradation, which may be the possible reason for the increase of TOF with battery degradation.

4.2. Ultrasonic Results for the Abusive Test

As mentioned in Section 3, an overcharge test (up to 5 V) was forced onto battery A1 after a period of cycling test (210 cycles) to investigate the feasibility of the ultrasonic signal under a more severe abusive condition than the previous one. The voltage and current for battery A1 during the overcharge process are shown in Figure 7a. Figure 7b shows the battery surface temperature and temperature change rate. Battery A1 was charged using a constant current of 0.5C (0.9 A) to 5 V, then maintained at 5 V. During the constant current charging period, the temperature varied between 45 °C and 50 °C, but started to rise when the voltage exceeded 4.709 V. After that, both the voltage and temperature increased sharply. During the constant voltage charging period (holding at 5 V), the current did not keep decreasing as expected, but increased rapidly at 3.697 h. At the same time, swelling of this battery was visually observed. The temperature at 3.697 h was 62.19 °C. For safety concerns, the test was eventually stopped. At that time, the current of the battery reached 1C (1.8 A).

Figure 7c shows the top view of the battery before and after overcharge. Figure 7d shows the X-ray images (the side view) of the battery. The thickness of the swollen battery increased as compared with the battery before overcharge. In addition, there was electrode ruffling after the battery was overcharged. This result is different from the situation before overcharge which shows the aligned electrode layers. The changes in the thickness and the interlayers are mainly attributed to gas generation under overcharge. Some gases, such as carbon dioxide (CO_2) and methane (CH_4), have been identified in batteries operating under overcharge conditions [33]. As the result and a product of electrolyte decomposition, the gas species are beyond the scope of this paper.

The ultrasonic features during the overcharge test are shown in Figure 7e,f. Figure 7e shows that the TOF shifts towards the lower value before 2.5 h, which is in agreement with [26] (due to the change of SOC). The TOF begins to increase at about 2.8 h. One possible interpretation of this increase is that the TOF has already indicated a tiny amount of gas generated inside the battery. In addition, an increase of the thickness of battery layers may also increase the value of TOF. Then, the TOF increases sharply in a similar time range as the temperature increases. Subsequently, the TOF shows the same decreasing trend as the temperature when the test is stopped. Figure 7f shows the change in the PA during the overcharge process. Similar to the cycling test results, the PA does not show a stable indication of the overcharge-induced battery failure.

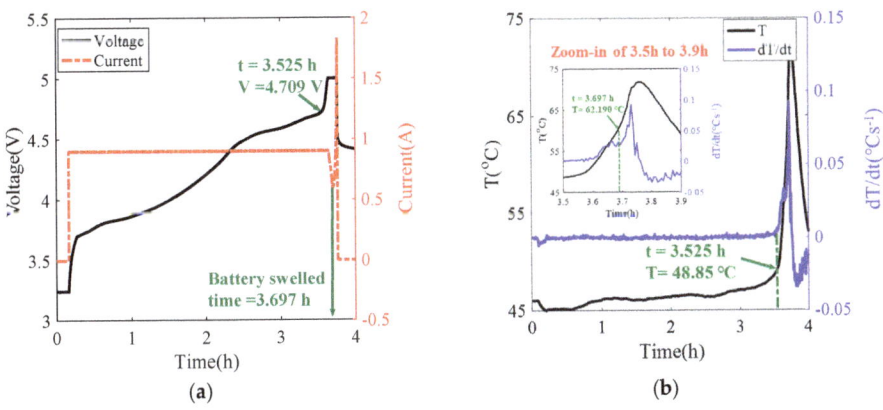

Figure 7. *Cont.*

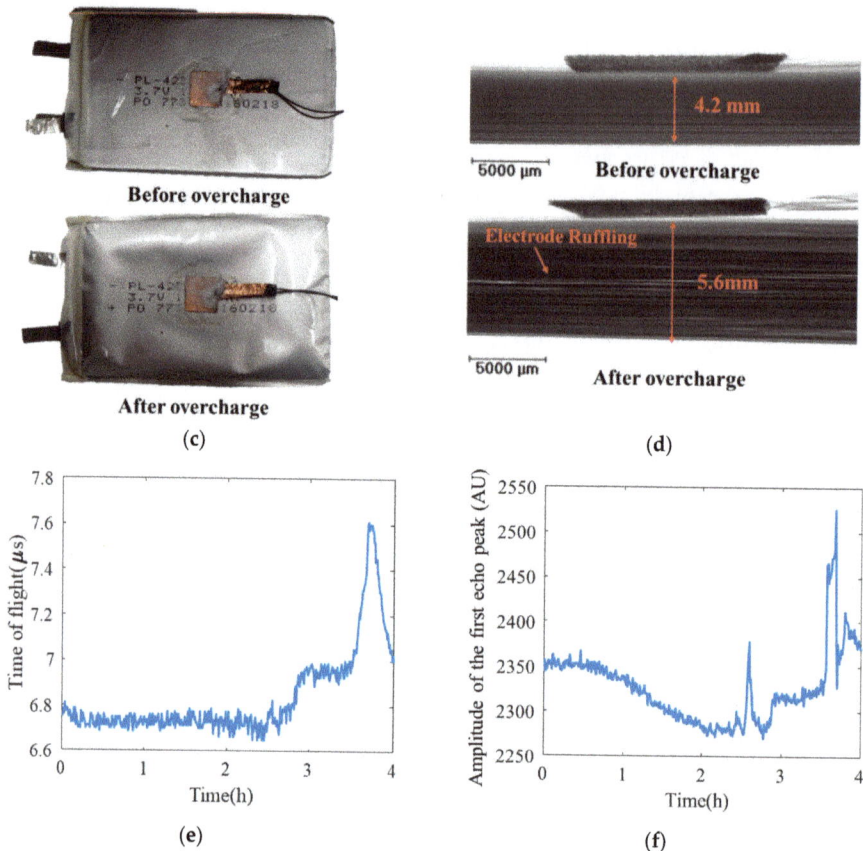

Figure 7. Overcharge test results for battery A1: (**a**) voltage and current profiles, (**b**) temperature profile, (**c**) photos of the battery (top view), (**d**) X-ray images of the battery (side view), (**e**) evolution of TOF, and (**f**) evolution of PA.

5. Ultrasonic Health Monitoring for Lithium-Ion Batteries

As summarized in the ultrasonic test results, the TOF was found to be highly related to the underlying degradation and failure process for the batteries, and therefore can be used for battery health monitoring. However, evaluation of battery degradation and failure will be arbitrary if a unique sensor is used because one sensor can have large uncertainties. Other than the ultrasonic signal, temperature is indicative of the health status of a battery, especially for some catastrophic failures such as thermal runaway. Therefore, to enhance the reliability of battery health monitoring, a data fusion method was developed that fuses the temperature with the ultrasonic signal.

5.1. Methodology

Mahalanobis distance (MD) is a distance measure with multivariate data that determine the similarity between an unknown sample and a collection of known samples [34]. It is sensitive to changes in different parameters because it considers the correlation among parameters, and it eliminates the scale problem by normalizing different parameters [35]. Due to these advantages, the MD approach was used as the data fusion method in this study. The MD values were calculated based on the temperature and ultrasonic features and used as the health indicators (HIs).

Figure 8 shows a flowchart of the MD-based battery health monitoring method. A dataset from the pristine battery (i.e., the healthy battery) was used as the training data to construct the baseline and determine the testing thresholds. Then, the MD value of the data from the test battery was calculated and compared with the thresholds to determine the health of the test battery.

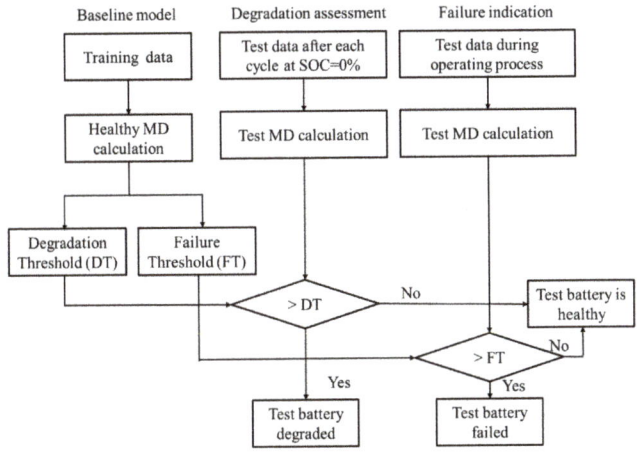

Figure 8. Developed battery health monitoring method.

The data collected from the healthy battery are denoted X_{ij}, where $i = 1, \ldots, m; j = 1, \ldots, n; m$ is the dimension of the feature vector; and n denotes the number of the observations. Each observation is normalized using the mean ($\overline{X}_i$) and the standard deviation (S_i) of the feature calculated from the healthy data. Thus, a parameter's normalized value is:

$$Z_{ij} = \frac{(X_{ij} - \overline{X}_i)}{S_i} \tag{4}$$

where

$$\overline{X}_i = \frac{1}{n}\sum_{j=1}^{n} X_{ij} \tag{5}$$

$$S_i = \sqrt{\frac{\sum_{j=1}^{n}(X_{ij} - \overline{X}_i)^2}{(n-1)}} \tag{6}$$

Then, the MD value for the healthy data is calculated by the following equation:

$$MD_j = Z_j^T C^{-1} Z_j \tag{7}$$

where C is the correlation matrix.

$$C = \frac{1}{(n-1)}\sum_{j=1}^{n} Z_j Z_j^T \tag{8}$$

Two types of test data were used to calculate the HIs for the purpose of health monitoring. To assess degradation, the HI was calculated based on the features at a specific state (the same SOC level) in order to eliminate the effect of SOC on the features. For the indication of early failure, the HI was obtained using the features, which were continuously monitored during the operating process, in order to provide an immediate response to a sudden failure. For different types of test data, different thresholds were established. For a battery, a degradation threshold is determined based on its degradation criterion. Generally, it is suggested to replace a battery if the capacity is less than 80% of its initial value [36]. Thus, the MD value, which corresponded to SOH = 80%, was set as the threshold for the battery degradation assessment.

Determination of the early failure threshold is an important step in order to have advanced warning of failure. A fixed threshold is defined based on expert knowledge. However, this approach may not be able to indicate early failure when knowledge of failure is not available. Therefore, it is useful to implement the generalized probabilistic approach to determine the threshold. The MD values are always positive and do not follow a normal distribution. A Box–Cox transformation transforms the data with these kinds of characteristics into a normal distribution. Then the mean (μ) and standard deviation (σ) of the transformed data are obtained. Since the intervals ($-\infty$, $\mu + 3\sigma$) contain 99.9% of the MD data, the three standard deviation limit ($\mu + 3\sigma$) was chosen as the threshold for failure indication, which has been demonstrated for various electronic products [35,37,38].

The Box–Cox transformation is defined as follows [39]:

$$y(\lambda) = \begin{cases} \frac{(y^\lambda - 1)}{\lambda}, & \lambda \neq 0 \\ \ln(y), & \lambda = 0 \end{cases} \quad (9)$$

where $y = y_1, y_2, \ldots, y_n$ is the original MD value and $y(\lambda)$ is the transformed MD value. The power λ is obtained by maximizing the log-likelihood function.

$$f(y, \lambda) = -\frac{n}{2} \ln\left[\sum_{i=1}^{n} \frac{(y_i(\lambda) - \bar{y}(\lambda))^2}{n}\right] + (\lambda - 1) \sum_{i=1}^{n} \ln(y_i) \quad (10)$$

where

$$\bar{y}(\lambda) = \frac{1}{n} \sum_{i=1}^{n} y_i(\lambda) \quad (11)$$

The mean (μ) and standard deviations (σ) of the transformed healthy MD values are used to determine the failure threshold. The upper bound ($\mu + 3\sigma$) is used as a threshold for failure indication [38].

For a test battery, each observation is normalized using the mean and standard deviation from the healthy data. Then, the MD value for each observation is calculated using Equation (7) and transformed using Equation (9). When the transformed test MD values cross the thresholds, a failure is considered to occur.

5.2. Battery Health Monitoring Results

As shown in Section 4, battery A1 underwent a period of cycling test and an overcharge test, sequentially. It is used as a case study to evaluate the developed health monitoring method. Figure 9a shows a scatter plot of the TOF and temperature for battery A1. First, the baseline data correspond to the first 10 data points from the cycling test. They refer to 100% to 98% SOH and are all measured at 0% SOC. Then, the test data 1 are extracted from the subsequent cycling test. They are also measured at 0% SOC but used for degradation assessment. Finally, the test data 2 correspond to the data measured during the 5 V overcharge process after 210 cycles. They are used for failure indication. Here, since tests were conducted under the well-controlled lab conditions, the data are used without a denoising process. In practice, the measured data may contain noise, these data are denoised before calculating the MD values.

The MD values corresponding to the training (baseline) data and the test data 1 for degradation assessment are shown in Figure 9b. These MD values were transformed into normally distributed variables using Box–Cox. To set the degradation threshold (DT), a linear model between the SOH and TOF was built based on the training data. Then, the estimated TOF value (TOF_DT) corresponding to SOH = 80% was obtained based on the linear model. In addition, the average value of the temperatures

(T_DT) from the training data was calculated. The MD value between (TOF_DT and T_DT) and the baseline was calculated and set as the DT. In order to quantify the degree of degradation, the MD values were normalized using the DT value. The range of the normalized MD value is from zero to one. A larger MD value indicates greater degradation, i.e., one means the battery degrades severely and is not able to be used and zero means the battery is healthy. According to Figure 9b, the normalized MD value shows an increasing tendency with the increasing cycle number, which means the battery degrades with the increasing cycle number. The MD value does not reach the degradation threshold until the end of the cycling test. The battery capacity is 96% of the initial capacity at the end of the cycling test, which also indicates the battery does not reach its degradation criterion (80% of its initial capacity). Therefore, the degradation assessment based on the proposed HI is consistent with the capacity test result.

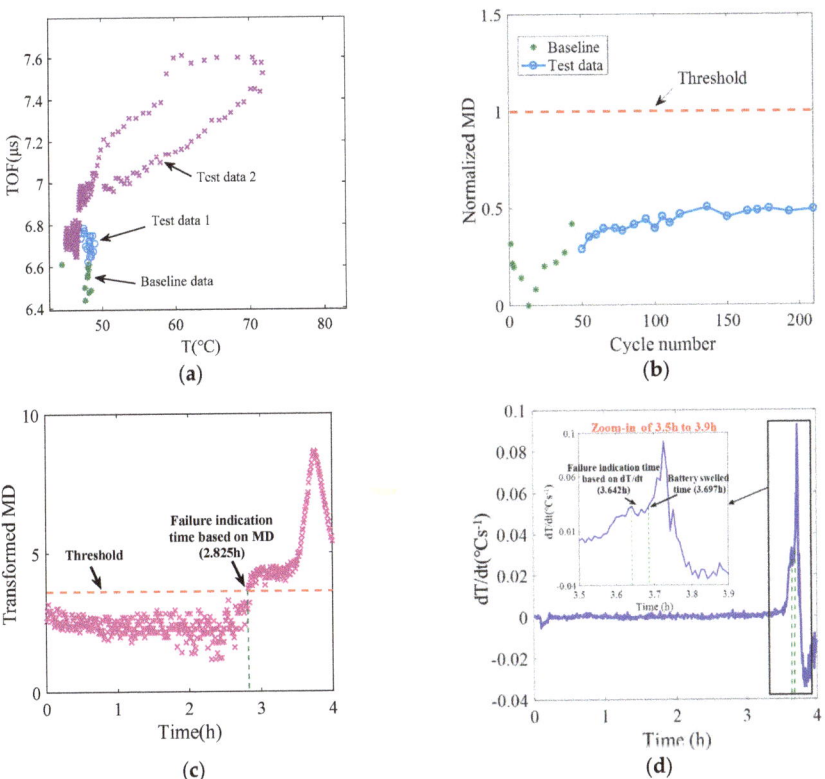

Figure 9. Health monitoring for battery A1: (**a**) scatter plot of the TOF and temperature, (**b**) degradation assessment result based on Mahalanobis distance (MD) values, (**c**) failure indication result based on MD values, and (**d**) failure indication result based on the temperature change rate.

The transformed MD values corresponding to test data 2 for failure indication are shown in Figure 9c. The failure threshold (FT) is plotted as the dashed line in Figure 9c. The temperature change rate (dT/dt) is plotted in Figure 9d. These two figures compare the indication time of failure by using the MD values and the temperature change rate, respectively. As shown in Figure 9c, when the MD value exceeded the failure threshold, the failure was indicated at 2.825 h. This time corresponds to the battery voltage at 4.564 V and the surface temperature at 46.86 °C. However, if the failure time is determined only in terms of the first peak of dT/dt, the battery failure was only identified at 3.642 h, which corresponds to close to 5 V voltage and a temperature of 55.65 °C. After that point, the temperature increased sharply, which means there was little time to prevent the battery from swelling. Therefore, the MD-based HI provides earlier failure indication time and a longer time margin for countermeasures to prevent further catastrophic failure as compared with monitoring only the temperature.

6. Conclusions

This paper developed an ultrasonic health monitoring method for lithium-ion batteries. To begin with, the feasibility of using ultrasonic sensing to probe the health status of a lithium-ion battery was demonstrated through battery cycling tests and an overcharge abusive test (up to 5 V). The ultrasonic results from the cycling test showed a strong dependence between the ultrasonic TOF and battery degradation. The overcharge test results showed that the ultrasonic TOF is sensitive to the battery swelling, which offers the potential for battery failure indication.

More importantly, an effective method for battery health monitoring was developed, which offers a significant improvement over the state-of-the-art ultrasonic techniques in terms of providing an early indication of sudden failure. A new data fusion MD-based health indicator was constructed by integrating the data from the temperature sensor and the ultrasonic transducer. The effectiveness of the health indicator was verified by the case study of battery A1, which underwent a cycling test followed by an overcharge test. The cycling test showed that the MD value increases with the battery aging. This means it can be used to determine battery degradation straightforwardly without the information on the battery's discharge capacity. The overcharge test (up to 5 V) showed that the health indicator can indicate battery failure 0.872 h ahead of battery swelling and 0.817 h earlier than the temperature-based method. In comparison to the temperature-based method, the developed method provides earlier warning of catastrophic failure and a longer time margin for failure prevention.

The developed ultrasonic health monitoring method can assess battery degradation as well as indicate battery sudden failure without needing the battery to be fully charged/discharged. This health monitoring method can be applied to batteries during their operation by integrating simple and small equipment (a pulser-receiver module and a piezoelectric transducer) into the existing battery management system. Similar techniques can be extended and applied to batteries with different chemistries, scales, and form factors. In future research, we will investigate the feasibility of using an ultrasonic inspection method to identify different aging mechanisms of the lithium-ion batteries.

Author Contributions: Investigation, M.P.; methodology, Y.W. (Yi Wu); writing—original draft, Y.W. (Yi Wu); writing—review and editing, Y.W. (Youren Wang) and M.P.; supervision, M.P.; funding acquisition, W.K.C.Y.

Funding: This work was supported by the National Natural Science Foundation of China (grant number 61473242); funding for Outstanding Doctoral Dissertation in NUAA (grant number BCXJ14-04); and the Fundamental Research Funds for the Central Universities and Funding of Jiangsu Innovation Program for Graduate Education (grant number KYLX_0251).

Acknowledgments: The authors would like to thank the more than 150 companies and organizations that support research activities at the Center for Advanced Life Cycle Engineering (CALCE) at the University of Maryland.

Conflicts of Interest: The authors declare no conflict of interest.

References

1. Arora, P. Capacity fade mechanisms and side reactions in lithium-ion batteries. *J. Electrochem. Soc.* **1998**, *145*, 3647. [CrossRef]
2. Kim, G.H.; Smith, K.; Ireland, J.; Pesaran, A. Fail-safe design for large capacity lithium-ion battery systems. *J. Power Sources* **2012**, *210*, 243–253. [CrossRef]
3. Wu, Y.; Saxena, S.; Xing, Y.; Wang, Y.; Li, C.; Yung, W.K.C.; Pecht, M. Analysis of manufacturing-induced defects and structural deformations in lithium-ion batteries using computed tomography. *Energies* **2018**, *11*, 925. [CrossRef]
4. Williard, N.; He, W.; Hendricks, C.; Pecht, M. Lessons learned from the 787 Dreamliner issue on lithium-ion battery reliability. *Energies* **2013**, *6*, 4682–4695. [CrossRef]
5. Hendricks, C.; Williard, N.; Mathew, S.; Pecht, M. A failure modes, mechanisms, and effects analysis (FMMEA) of lithium-ion batteries. *J. Power Sources* **2015**, *297*, 113–120. [CrossRef]

6. Abada, S.; Marlair, G.; Lecocq, A.; Petit, M.; Sauvant-Moynot, V.; Huet, F. Safety focused modeling of lithium-ion batteries: A review. *J. Power Sources* **2016**, *306*, 178–192. [CrossRef]
7. El Mejdoubi, A.; Oukaour, A.; Chaoui, H.; Gualous, H.; Sabor, J.; Slamani, Y. State-of-charge and state-of-health lithium-ion batteries' diagnosis according to surface temperature variation. *IEEE Trans. Ind. Electron.* **2016**, *63*, 2391–2402. [CrossRef]
8. Berecibar, M.; Gandiaga, I.; Villarreal, I.; Omar, N.; Van Mierlo, J.; Van den Bossche, P. Critical review of state of health estimation methods of Li-ion batteries for real applications. *Renew. Sustain. Energy Rev.* **2016**, *56*, 572–587. [CrossRef]
9. Lipu, M.S.H.; Hannan, M.A.; Hussain, A.; Hoque, M.M.; Ker, P.J.; Saad, M.H.M.; Ayob, A. A review of state of health and remaining useful life estimation methods for lithium-ion battery in electric vehicles: Challenges and recommendations. *J. Clean. Prod.* **2018**, *205*, 115–133. [CrossRef]
10. Ahwiadi, M.; Wang, W. An enhanced mutated particle filter technique for system state estimation and battery life prediction. *IEEE Trans. Instrum. Meas.* **2019**, *68*, 923–935. [CrossRef]
11. Liu, Z.; Sun, G.; Bu, S.; Han, J.; Tang, X.; Pecht, M. Particle learning framework for estimating the remaining useful life of lithium-ion batteries. *IEEE Trans. Instrum. Meas.* **2017**, *66*, 280–293. [CrossRef]
12. Liu, D.; Song, Y.; Li, L.; Liao, H.; Peng, Y. On-line life cycle health assessment for lithium-ion battery in electric vehicles. *J. Clean. Prod.* **2018**, *199*, 1050–1065. [CrossRef]
13. Xing, Y.; Ma, E.W.M.; Tsui, K.; Pecht, M. An ensemble model for predicting the remaining useful performance of lithium-ion batteries. *Microelectron. Reliab.* **2013**, *53*, 811–820. [CrossRef]
14. Novais, S.; Nascimento, M.; Grande, L.; Domingues, M.F.; Antunes, P.; Alberto, N.; Leitão, C.; Oliveira, R.; Koch, S.; Kim, G.T.; et al. Internal and external temperature monitoring of a Li-ion battery with fiber Bragg grating sensors. *Sensors* **2016**, *16*, 1394. [CrossRef] [PubMed]
15. Raghavan, A.; Kiesel, P.; Sommer, L.W.; Schwartz, J.; Lochbaum, A.; Hegyi, A.; Schuh, A.; Arakaki, K.; Saha, B.; Ganguli, A.; et al. Embedded fiber-optic sensing for accurate internal monitoring of cell state in advanced battery management systems part 1: Cell embedding method and performance. *J. Power Sources* **2017**, *341*, 466–473. [CrossRef]
16. Ganguli, A.; Saha, B.; Raghavan, A.; Kiesel, P.; Arakaki, K.; Schuh, A.; Schwartz, J.; Hegyi, A.; Sommer, L.W.; Lochbaum, A.; et al. Embedded fiber-optic sensing for accurate internal monitoring of cell state in advanced battery management systems part 2: Internal cell signals and utility for state estimation. *J. Power Sources* **2017**, *341*, 474–482. [CrossRef]
17. Lee, C.-Y.; Lee, S.-J.; Hung, Y.-M.; Hsieh, C.-T.; Chang, Y.-M.; Huang, Y.-T.; Lin, J.-T. Integrated microsensor for real-time microscopic monitoring of local temperature, voltage and current inside lithium ion battery. *Sens. Actuators A Phys.* **2017**, *253*, 59–68. [CrossRef]
18. Giurgiutiu, V. *Structural Health Monitoring: With Piezoelectric Wafer Active Sensors*; Elsevier: Amsterdam, The Netherlands, 2007.
19. Schmerr, L.W. *Fundamentals of Ultrasonic Nondestructive Evaluation*; Springer: New York, NY, USA, 2016.
20. Inspection of Prismatic Lithium Ion Batteries for Consumer Electronics. Available online: http://ultrangroup.com/applications/inspection-of-prismatic-lithium-ion-batteries-for-consumer-electronics/ (accessed on 12 March 2019).
21. Sood, B.; Osterman, M.; Pecht, M. Health monitoring of lithium-ion batteries. In Proceedings of the 2013 IEEE Symposium on Product Compliance Engineering (ISPCE), Austin, TX, USA, 7–9 October 2013; pp. 1–6.
22. Li, H.; Zhou, Z.; Li, H.; Zhou, Z. Numerical simulation and experimental study of fluid-solid coupling-based air-coupled ultrasonic detection of stomata defect of lithium-ion battery. *Sensors* **2019**, *19*, 2391. [CrossRef] [PubMed]
23. Ladpli, P.; Nardari, R.; Kopsaftopoulos, F.; Wang, Y.; Chang, F.K. Design of multifunctional structural batteries with health monitoring capabilities. In Proceedings of the 8th European Workshop on Structural Health Monitoring, Bilbao, Spain, 5–8 July 2016; pp. 1–13.
24. Ladpli, P.; Kopsaftopoulos, F.; Nardari, R.; Chang, F.K. Battery charge and health state monitoring via ultrasonic guided-wave-based methods using built-in piezoelectric transducers. In Proceedings of the Smart Materials and Nondestructive Evaluation for Energy Systems, Portland, OR, USA, 25–29 March 2017; p. 1017108.
25. Ladpli, P.; Kopsaftopoulos, F.; Chang, F.-K. Estimating state of charge and health of lithium-ion batteries with guided waves using built-in piezoelectric sensors/actuators. *J. Power Sources* **2018**, *384*, 342–354. [CrossRef]

26. Hsieh, A.G.; Bhadra, S.; Hertzberg, B.; Gjeltema, P.J.; Goy, A.; Fleischer, J.W.; Steingart, D. Electrochemical-acoustic time of flight: In operando correlation of physical dynamics with battery charge and health. *Energy Environ. Sci.* **2015**, *8*, 1569–1577. [CrossRef]
27. Gold, L.; Bach, T.; Virsik, W.; Schmitt, A.; Müller, J.; Staab, T.E.M.; Sextl, G. Probing lithium-ion batteries' state-of-charge using ultrasonic transmission—Concept and laboratory testing. *J. Power Sources* **2017**, *343*, 536–544. [CrossRef]
28. Davies, G.; Knehr, K.W.; Van Tassell, B.; Hodson, T.; Biswas, S.; Hsieh, A.G.; Steingart, D.A. State of charge and state of health estimation using electrochemical acoustic time of flight analysis. *J. Electrochem. Soc.* **2017**, *164*, A2746–A2755. [CrossRef]
29. Choi, J.W.; Aurbach, D. Promise and reality of post-lithium-ion batteries with high energy densities. *Nat. Rev. Mater.* **2016**, *1*, 16013. [CrossRef]
30. Olivares, B.E.; Cerda Munoz, M.A.; Orchard, M.E.; Silva, J.F. Particle-filtering-based prognosis framework for energy storage devices with a statistical characterization of state-of-health regeneration phenomena. *IEEE Trans. Instrum. Meas.* **2013**, *62*, 364–376. [CrossRef]
31. Swallow, J.G.; Woodford, W.H.; McGrogan, F.P.; Ferralis, N.; Chiang, Y.-M.; Van Vliet, K.J. Effect of electrochemical charging on elastoplastic properties and fracture toughness of Li_xCoO_2. *J. Electrochem. Soc.* **2014**, *161*, F3084–F3090. [CrossRef]
32. Jameson, N.J.; Azarian, M.H.; Pecht, M. Impedance-based condition monitoring for insulation systems used in low-voltage electromagnetic coils. *IEEE Trans. Ind. Electron.* **2017**, *64*, 3748–3757. [CrossRef]
33. Ohsaki, T.; Kishi, T.; Kuboki, T.; Takami, N.; Shimura, N.; Sato, Y.; Sekino, M.; Satoh, A. Overcharge reaction of lithium-ion batteries. *J. Power Sources* **2005**, *146*, 97–100. [CrossRef]
34. Wang, Y.; Miao, Q.; Ma, E.W.M.; Tsui, K.L.; Pecht, M.G. Online anomaly detection for hard disk drives based on Mahalanobis distance. *IEEE Trans. Reliab.* **2013**, *62*, 136–145. [CrossRef]
35. Patil, N.; Das, D.; Pecht, M. Anomaly detection for IGBTs using Mahalanobis distance. *Microelectron. Reliab.* **2015**, *55*, 1054–1059. [CrossRef]
36. Berecibar, M.; Garmendia, M.; Gandiaga, I.; Crego, J.; Villarreal, I. State of health estimation algorithm of $LiFePO_4$ battery packs based on differential voltage curves for battery management system application. *Energy* **2016**, *103*, 784–796. [CrossRef]
37. Kumar, S.; Chow, T.W.S.; Pecht, M. Approach to fault identification for electronic products using Mahalanobis distance. *IEEE Trans. Instrum. Meas.* **2010**, *59*, 2055–2064. [CrossRef]
38. Jin, X.; Wang, Y.; Chow, T.W.S.; Sun, Y. MD-based approaches for system health monitoring: A review. *IET Sci. Meas. Technol.* **2017**, *11*, 371–379. [CrossRef]
39. Box, G.E.P.; Box, G.E.P.; Cox, D.R. An analysis of transformations. *J. R. Stat. Soc. Ser. B* **1964**, *26*, 211–252. [CrossRef]

© 2019 by the authors. Licensee MDPI, Basel, Switzerland. This article is an open access article distributed under the terms and conditions of the Creative Commons Attribution (CC BY) license (http://creativecommons.org/licenses/by/4.0/).

Article

Battery Charger Based on a Resonant Converter for High-Power LiFePO$_4$ Batteries

Christian Brañas [1,*], Juan C. Viera [2], Francisco J. Azcondo [1], Rosario Casanueva [1], Manuela Gonzalez [2] and Francisco J. Díaz [1]

1. Electronics Technology, Systems and Automation Engineering Department, University of Cantabria, 39005 Santander, Spain; azcondof@unican.es (F.J.A.); casanuer@unican.es (R.C.); diazrf@unican.es (F.J.D.)
2. Department of Electrical Engineering, University of Oviedo, Campus de Gijón, Módulo 3, 33204 Gijón, Spain; viera@uniovi.es (J.C.V.); mgonzalez@uniovi.es (M.G.)
* Correspondence: branasc@unican.es; Tel.: +34-942-200-873

Citation: Brañas, C.; Viera, J.C.; Azcondo, F.J.; Casanueva, R.; Gonzalez, M.; Díaz, F.J. Battery Charger Based on a Resonant Converter for High-Power LiFePO$_4$ Batteries. *Electronics* **2021**, *10*, 266. https://doi.org/10.3390/electronics10030266

Received: 21 December 2020
Accepted: 21 January 2021
Published: 23 January 2021

Publisher's Note: MDPI stays neutral with regard to jurisdictional claims in published maps and institutional affiliations.

Copyright: © 2021 by the authors. Licensee MDPI, Basel, Switzerland. This article is an open access article distributed under the terms and conditions of the Creative Commons Attribution (CC BY) license (https://creativecommons.org/licenses/by/4.0/).

Abstract: A new battery charger, based on a multiphase resonant converter, for a high-capacity 48 V LiFePO$_4$ lithium-ion battery is presented. LiFePO$_4$ batteries are among the most widely used today and offer high energy efficiency, high safety performance, very good temperature behavior, and a long cycle life. An accurate control of the charging current is necessary to preserve the battery health. The design of the charger is presented in a tight correlation with a battery model based on experimental data obtained at the laboratory. With the aim of reducing conduction losses, the general analysis of the inverter stage obtained from the parallel connection of N class D LC_pC_s resonant inverters is carried out. The study provides criteria for proper selection of the transistors and diodes as well as the value of the DC-link voltage. The effect of the leakage inductance of the transformer on the resonant circuit is also evaluated, and a design solution to cancel it is proposed. The output stage is based on a multi-winding current-doubler rectifier. The converter is designed to operate in open-loop operation as an input voltage-dependent current source, but in closed-loop operation, it behaves as a voltage source with an inherent maximum output current limitation, which provides high reliability throughout the whole charging process. The curve of efficiency of the proposed charger exhibits a wide flat zone that includes light load conditions.

Keywords: lithium-ion battery; battery modeling; battery chargers; power supplies; resonant inverters; phase control

1. Introduction

Lithium iron phosphate (LiFePO$_4$) batteries have a great electrochemical performance and a good thermal stability, which makes them safer and more robust. This lithium-based technology exhibits a very low internal resistance offering a high current rating. Their cycle life is significantly longer compared to other technologies [1,2]. The applications of LiFePO$_4$ batteries are, among others, for storage systems in renewable energy facilities, powering electric vehicles and uninterruptible power supplies (UPS) in data centers, telecommunications, and hospitals. A battery model is an important tool for designing the charger allowing the study of the dynamic response of the battery-charger system along the whole charging process, wherein the converter load, i.e., the equivalent resistance of the battery, varies from almost short-circuit to open-circuit values. Most of the battery models are aimed to improve the battery management system (BMS) performance, providing information about important parameters of the battery such as the state of charge (SOC) [3]. The estimation of the battery SOC and power capacity is usually solved by applying three methods, i.e., the look-up table method, the model-based method, and the artificial intelligence method [4–7]. In addition to that, the BMS is responsible for ensuring the battery operation within safety margins of temperature and sets the overvoltage and under voltage protection limits. In this work, the battery modeling is presented in a tight correlation with the battery charger design.

The technology of resonant converters is chosen to implement the proposed battery charger. The advantages of the resonant conversion of energy, such as high frequency of operation, sinusoidal waveforms, and low switching losses are well known [8]. Among all possible configurations of resonant converters, the series resonant converter and the *LLC* converter have been widely used [9–12]. Usually, the converter is designed to operate as a voltage source with some kind of control to limit the charging current. In this work, the converter is designed as a voltage-dependent current source. In this approach, the circuit presents an inherent maximum current limitation, which is a safer operation mode. The LiFePO$_4$ technology reaches current rates as high as hundreds of amps. In circuit design for high-current applications, conduction losses are a major design limitation [13,14]. In high-current resonant converters, increasing the dc-link voltage, V_{dc}, and using a step-down transformer ($n > 1$) reduces the amplitude of the resonant currents in the inverter stage, minimizing the conduction loss in transistors and resonant inductors. New Wide Band Gap (WBG) devices enable the operation at an 800 V to 1700 V dc-link voltages range [15]. WBG devices achieve high performance at high current levels with important simplifications in the power circuit. However, the cost of WBG devices limits their use for certain applications.

In this work, a generalized design method aimed at minimizing the conduction loss is presented for multiphase resonant converters [16]. The number of parallel branches and therefore phases, N, in the inverter stage is calculated according to the maximum output power and the expected efficiency. This alternative offers another degree of freedom for achieving efficiencies higher than 90% even at relatively low values of V_{dc} and using low-cost transistors. Moreover, the multi-phase structure makes it possible to regulate the charging current at constant switching frequency by shifting the phase of the output voltages of each class D section of the inverter.

This paper is organized as follows: After the introduction, Section 2 describes the charging profile of the target LiFePO$_4$ battery, which is oriented to obtain a fast charge without reducing its lifetime. The battery model is presented in Section 3. The analysis of the proposed charger and main design equations are developed in Section 4. The efficiency of the charger is studied in Section 5. A detailed step-by-step design sequence of the proposed charger is explained in Section 6. In Sections 7 and 8, the results obtained for the modeling of the battery and experimental waveforms to verify the performance of the prototype are presented, ending with a discussion about Si vs. SiC solutions and concluding remarks.

2. Charging Method

The main characteristics of the commercial 48NPFC50 LiFePO$_4$ battery (Narada Power Source Co., Ltd., Hangzhou, China) [17] used in this work are 48 V nominal voltage and 50 Ah nominal capacity (C_n) i.e., 2.4 kWh of power capacity. The battery consists of fifteen (N_s = 15) stacked cells in series and incorporates a BMS that guarantees the right balance-of-charge of all cells. Thus, the voltage across each cell is assumed identical to any other. The battery charger is designed to meet all operational limits settled by the BMS.

The charging protocol recommended for LiFePO$_4$ batteries is the well-known [18] constant current (CC)–constant voltage (CV) method (i.e., CC–CV). During the CC stage, the battery is charged at the maximum current rate, which depends on the battery capacity and technology. Once the battery voltage reaches its maximum charging voltage specified in the battery data sheet, the CV stage begins. At this point, the power drawn from the charger is the maximum, which happens at 90% of the *SOC* approximately. During the CV stage, the charging current diminishes. Three experimental charging profiles are carried on at the battery laboratory facility shown in Figure 1. They are evaluated at room temperature (25 °C) using the battery test equipment PEC SBT-10050 (PEC, Leuven, Belgium) and taking into account that the battery is fully discharged as the initial condition.

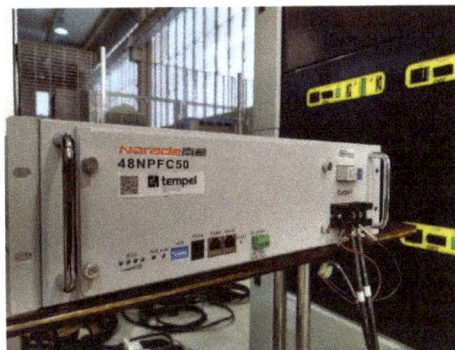

Figure 1. (**Left**) Laboratory facility for battery testing. (**Right**) Battery connected to SBT-10050 test equipment.

Those profiles correspond to the battery charge at current rates equal to $C_n/5$, $C_n/2$, and C_n during the CC stage. The results are shown in Figure 2.

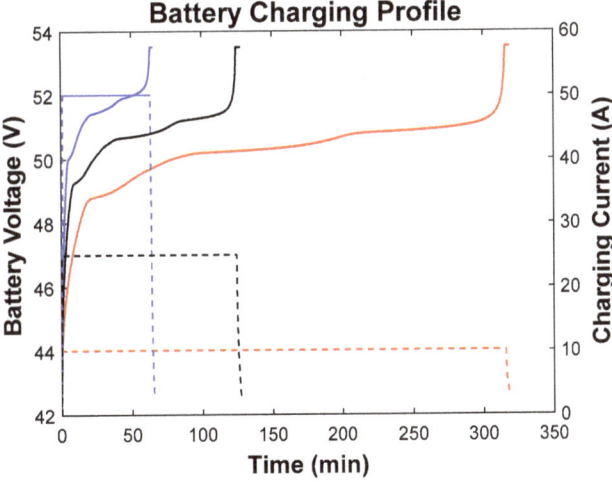

Figure 2. Experimental charging profiles at 10, 25, and 50 A for a 48NPFC50 LiFePO$_4$ battery

The temperature is observed by the BMS during the whole charging process, and it implements the corresponding protection (maximum value 55 °C for charging) to prevent the battery aging. Electro-thermal models for studying the temperature of a lithium-ion cell as a function of the charging/discharging current have been reported in [19,20]. The user manual recommends a conservative value, $C_n/5$, for the charging current rate; however, LiFePO$_4$ technology tolerates fast-charging protocols [21–24]. In this work, in order to shorten the charging time, a maximum charging current rate of 20 A (approximately $C_n/2$) is chosen the charger design. According to the experimental characterization of the battery, charging at $C_n/2$ keeps the temperature of the battery well below 55 °C.

3. Battery Model

Although the LiFePO$_4$ cell is a complex physical system with several variables involved, a good trade-off among simplicity, accuracy, and insight information is obtained with the electrical parameters-based models [25], as shown in Figure 3. The single cell model is generalized by affecting all parameters by the total number of cells, N_s, under the assumption that all cells are identical, as shown in Figure 3.

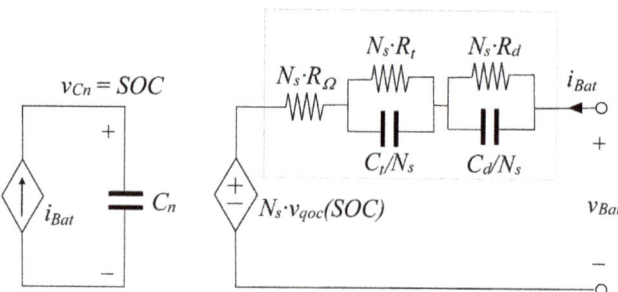

Figure 3. Battery model considering N_s stacked cells in series.

The state of charge (SOC) [26] of the battery is defined as the ratio of the battery charge, Q, to the nominal capacity, C_n.

$$SOC = \frac{Q}{C_n} \cdot 100\% \qquad (1)$$

The model calculates the SOC [26], integrating the battery current-dependent current source, i_{bat}, which charges/discharges the capacitor C_n. The SOC is equal to the voltage across the capacitor C_n, v_{Cn} varying from zero to one corresponding to exhausted to fully charged battery, respectively.

$$SOC(t) = SOC(t_o) + \frac{1}{C_n} \int_{t_o}^{t} i_{bat}(t) dt \qquad (2)$$

The voltage-controlled voltage source, $N_s v_{qoc}(SOC)$, dependent on the voltage v_{Cn}, represents the quasi-open-circuit battery voltage, where v_{qoc}, is the quasi-open-circuit voltage across one single cell. The experimental measurement of v_{qoc} as a function of the SOC is a time-consuming task because it should be obtained while keeping the cell in electrochemical equilibrium [27], charging and discharging the cell at a very low current rate. From the experimental study of one single cell, the v_{qoc} as a function of the SOC was obtained by charging and discharging the cell at $C_n/50$. This test required 100 h. The result is shown in Figure 4.

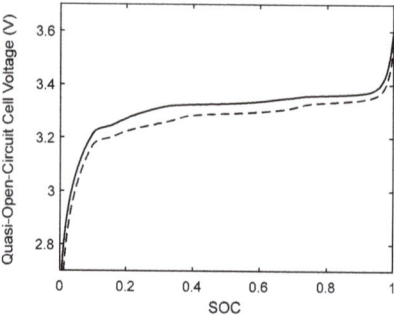

Figure 4. Quasi-open-circuit voltage of the cell as a function of the state of charge (SOC) obtained at $C_n/50$ for a complete charge/discharge cycle. Solid line: Charge trajectory. Dashed line: Discharge trajectory.

As it is observed in Figure 4, the quasi-open-circuit cell voltage, v_{qoc}, incorporates the effect of the voltage hysteresis caused by the battery structure [27]. The maximum hysteresis is about ≈ 40 mV within the 30% SOC region, and the average is ≈ 20 mV within the 40% to 80% SOC region. The experimental test results show a cell capacity $C_n = 50$ Ah, which is represented in the model by a capacitance $C_n = 180,000$ F.

The electrolyte and electrode resistance are modeled by R_Ω. In addition to that, the model also includes two time constants, which are modeled by networks $R_t C_t$ and $R_d C_d$. The time constant $R_t C_t$ is associated to chemical reactions and charge transportation phenomenon in the electrodes. This time constant is within the range from milliseconds to a few seconds. In contrast, the time constant $R_d C_d$ governs the mass diffusion in the electrolyte and electrodes and is within the tens of seconds range [27]. From the point of view of the battery charger design, the electrical parameters of the battery at the end of the CC stage are of interest. At this point, the power supplied by the charger is the maximum. For a given SOC, the battery model can be simplified to a resistance, r_{Bat}, in series with a voltage source equal to the quasi-open-circuit voltage $N_s V_{qoc}$. Assuming the battery is in steady state, r_{bat} is obtained from the model shown in Figure 3 as

$$r_{Bat} = N_s \cdot (R_\Omega + R_t + R_d). \qquad (3)$$

The specific values R_t and R_d for a given SOC should be obtained from the dynamic study of the battery, once the time constants associated with transport and diffusion phenomena were obtained. Finally, the battery voltage is obtained as:

$$V_{Bat} = N_s V_{qoc} + I_{Bat} r_{Bat}. \qquad (4)$$

4. Multiphase $LC_p C_s$ Resonant Converter

The proposed battery charger is a multiphase resonant converter. The general form of the circuit is shown in Figure 5, where the battery is modeled in steady state by its internal impedance, r_{Bat}, in series with the quasi-open-circuit battery voltage $N_s V_{qoc}$.

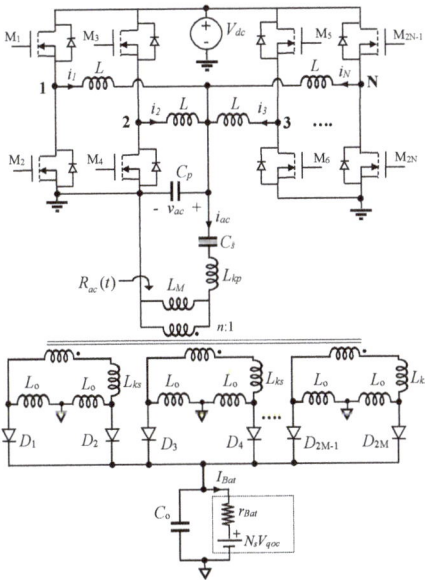

Figure 5. General architecture of the battery charger based on an N-phase $LC_p C_s$ resonant inverter with an M-winding current-doubler rectifier as output stage. Multiple configurations are possible according to the N and M values.

The AC side is a multiphase resonant inverter, which consists of N paralleled LC_pC_s class D sections [16,28]. Among the possible configurations of the resonant network, the configuration LC_pC_s of the LCC family is chosen to achieve a current source behavior while preserving the zero voltage switching (ZVS) mode of transistors [8,29]. Unlike the LLC converter, the proposed LC_pC_s does not require a gapped-core transformer [30], so the magnetizing inductance, L_M, is high enough to neglect its impact in the later analysis.

The DC side consists of an M-winding current multiplier, which is derived from the parallel connection of an M current-doubler rectifier [31,32]. The low output voltage of this application recommends the use of Schottky diodes without any control circuit in the secondary side, which is a simplification in comparison to solutions based on synchronous rectification (SR).

4.1. Resonant Inverter Stage

The converter is analyzed considering the general case, where each midpoint voltage v_i of all class D sections has associated a phase-angle $\Psi_0, \Psi_1, \ldots, \Psi_{i-1}$. To illustrate this assumption, the midpoint voltages, v_i, are shown in Figure 6.

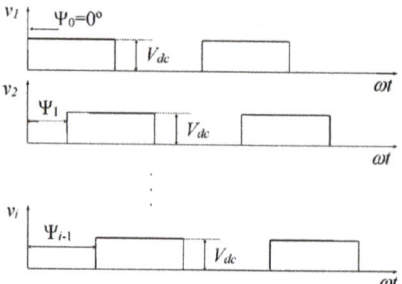

Figure 6. Output voltages of each inverter section obtained in the midpoint of the transistors leg.

Using the fundamental approximation, the input voltages, v_i, are represented with the exponential form given in (5),

$$\mathbf{V_i} = \frac{2V_{dc}}{\pi} \cdot e^{-j\Psi_{i-1}}, \tag{5}$$

where $i \in [1, 2, \ldots, N]$ is the phase number. In steady state and using the low ripple approximation, the M-winding output rectifier is reduced to an equivalent impedance R_{ac} [8,29]. The resonant inverter stage is analyzed using the simplified circuit model shown in Figure 7.

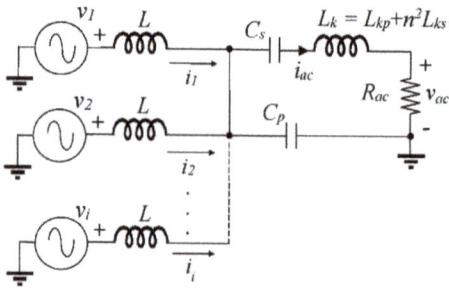

Figure 7. Simplified model using the fundamental approximation for circuit analysis purposes of the inverter stage.

The parallel parameters of the resonant inverter are defined in Table 1.

Table 1. Parameters of the LC_pC_s resonant inverter.

Parallel Resonant Frequency	Parallel Characteristic Impedance	Parallel Quality Factor
$\omega_p = \dfrac{1}{\sqrt{LC_p/N}}$	$Z_p = \omega_p L = \dfrac{N}{\omega_p C_p}$	$Q_p = \dfrac{NR_{ac}}{Z_p}$

4.1.1. AC Side Output Current

During the CC stage of the charging process, the converter provides an inherent current limitation, protecting the battery and extending its life. The current source behavior of the resonant converter is achieved by fixing the switching frequency at $\omega = \omega_p$, where ω_p is the parallel resonant frequency, as given in Table 1. Once the switching frequency is fixed at $\omega = \omega_p$, the output current, seen from the primary side of the transformer, i.e., through C_s, $\mathbf{I_{ac}}$, is calculated by (6).

$$\mathbf{I_{ac}} = -\frac{2V_{dc}}{\pi Z_p}\left\{\sum_{m=1}^{N}\sin\Psi_{m-1} + j\sum_{m=1}^{N}\cos\Psi_{m-1}\right\} \quad (6)$$

From (6), the current source behavior is verified, given that $\mathbf{I_{ac}}$ has no dependence on the load.

4.1.2. Switching Mode

The switching losses are minimized by ensuring the zero voltage switch (ZVS) on the primary side of the converter [8,29]. The ZVS mode requires sufficient phase-delay of the resonant current with respect to the input voltage. A high value of Q_p reduces the reactive energy in the resonant converter, which is beneficial from the point of view of reducing the conduction loss. However, some reactive energy must be accepted for ensuring the ZVS mode of all transistors. The complex form, $\mathbf{I_i}$, of each resonant current is given in (7) as a function of the angles $\Psi_0, \Psi_1, \ldots, \Psi_{N-1}$.

$$\mathbf{I_i} = \frac{2V_{dc}}{\pi Z_p} \times \left\{ \begin{array}{l} \dfrac{Q_p}{N}\sum_{m=1}^{N}\cos\Psi_{m-1} - \left(\dfrac{C_p}{NC_s} - \dfrac{L_k}{L}\right)\sum_{m=1}^{N}\sin\Psi_{m-1} - \sin\Psi_{i-1} \\[1em] -j\left[\dfrac{Q_p}{N}\sum_{m=1}^{N}\sin\Psi_{m-1} + \left(\dfrac{C_p}{NC_s} - \dfrac{L_k}{L}\right)\sum_{m=1}^{N}\cos\Psi_{m-1} + \cos\Psi_{i-1}\right] \end{array} \right\} \quad (7)$$

In order to determine the power factor angle, ϕ_i, of each transistor's leg, the input impedance $\mathbf{Z_i} = \mathbf{V_i}/\mathbf{I_i}$, of each phase is calculated. The power factor angle, ϕ_i is obtained using $\phi_i = \mathrm{angle}(\mathbf{Z_i})$ as a function of the control angles, $\Psi_0, \Psi_1, \ldots, \Psi_{N-1}$, the number of phases, N, and the quality factor, Q_p. Upon substitution of $\Psi_0 = \Psi_1 = \ldots \Psi_{N-1} = 0°$ in (5) and (7), ϕ_i at the maximum output current is obtained:

$$\phi_i = \arctan\left(\frac{1 + \frac{C_p}{C_s} - N\frac{L_k}{L}}{Q_p}\right). \quad (8)$$

From (8), it can be observed that the effect of leakage inductance referred to the primary side of the transformer, L_k, is more significant for high-power as well as for high-frequency designs, where the value of inductance of the resonant circuit, L, is usually low, and a high value of leakage inductance could produce the loss of the ZVS mode. However, the series disposition of L_k and C_s enables the cancelation of the L_k effect on the AC side by calculating C_s to achieve, at the switching frequency, the series resonance with L_k. According to this proposal, C_s is obtained from (9).

$$C_s = \frac{L}{NL_k}C_p \quad (9)$$

With the cancellation of the L_k effect, the value of the power factor angle, ϕ_i, depends essentially on the value of the quality factor Q_p, which is set during the design process of the converter. The minimum value of power factor angle for achieving ZVS, ϕ_{zvs}, depends on the dead time, t_d, of the transistors' driver and the switching frequency ω_p [33].

$$\varphi_{zvs} = \frac{t_d \omega_p}{2\pi} \cdot 360° \tag{10}$$

As design criteria, a value of power factor angle $\phi_i = 2\phi_{zvs}$ is assumed at nominal conditions for achieving a reliable operation of the converter. This is the most restrictive design condition for operating in ZVS mode for the whole range of variation of the control angle Ψ. From (8) to (10), the value of the quality factor at nominal conditions, Q_{pN}, is obtained:

$$Q_{pN} = \frac{1}{\tan 2\varphi_{zvs}} \tag{11}$$

4.1.3. Variation of the Quality Factor and Transformer Turns Ratio

During the charging process, the equivalent impedance of the battery, R_{Bat}, changes depending on V_{Bat} and I_{Bat}, whose relationship is given by the charging profile of the battery, as shown in Figure 1. At the end of the CC stage, $V_{Bat} = V_{Bat(Max)}$ and the power supplied to the battery reaches the maximum, $P_{Bat} = V_{Bat(Max)} I_{Bat(Max)}$. The specifications of the point of maximum output power are used for defining the nominal value of the quality factor, Q_{pN}. Thus, during the CC stage of the charging profile, the converter works with a quality factor lower than the nominal one, which strengthens the inductive behavior of the resonant tank, assuring the ZVS mode.

During the CV stage, the reduction of the charging current leads to a significant increment in the equivalent resistance R_{Bat} and consequently, the reflected impedance on the AC side, R_{ac}, and the quality factor Q_p also increase. Assuming that $V_{Bat(Max)}$ is constant and working with (6), the quality factor as a function of Ψ is obtained in (12),

$$Q_p = \frac{n\pi^2 V_{Bat(Max)}}{2V_{dc}} \cdot \frac{N}{\sqrt{\left(\sum_{m=1}^{N} \sin \Psi_{m-1}\right)^2 + \left(\sum_{m=1}^{N} \cos \Psi_{m-1}\right)^2}} \tag{12}$$

The increment of Q_p, as a consequence of the reduction of the charging current during the CV stage could put at risk the ZVS mode of the transistors of the converter. However, it is beneficial from the point of view of achieving waveforms with low distortion and increases the converter efficiency. The nominal value of quality factor is obtained by evaluating (12) for $\Psi_0 = \Psi_1 = \Psi_{N-1} = 0°$.

$$Q_{pN} = \frac{n\pi^2 V_{Bat(Max)}}{2V_{dc}} \tag{13}$$

From (13) and (11), the transformer's turns ratio (n:1) can be obtained:

$$n = \frac{2V_{dc}}{\pi^2 V_{Bat(Max)} \tan 2\varphi_{zvs}}. \tag{14}$$

4.2. Output Current Multiplier

In order to analyze the output current multiplier stage, first, a single-winding current-doubler rectifier with an ideal transformer, as seen in Figure 8, is considered. The quasi-sinusoidal voltage v_{ac} at parallel capacitor C_p drives the current multiplier stage. The diodes D_1 and D_2 turn on alternatively according the positive or negative cycle of v_{ac}, respectively.

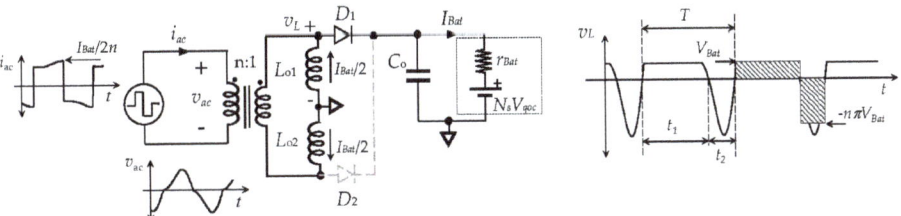

Figure 8. (**Left**) Current-doubler rectifier considering the positive semi-cycle of the drive voltage v_{ac}. (**Right**) Theoretical voltage waveform at the filter inductors.

The diodes conduction time, t_1, is obtained from the volts–seconds balance across the inductors. The areas are calculated according to the approximation shown in Figure 8 right.

$$t_1 = \frac{n\pi}{1+n\pi}T \tag{15}$$

The average current through each inductor, $L_{o1,2}$, is equal to one-half of the charging current I_{Bat}. The amplitude of the current ripple in each inductor is determined by

$$\Delta i_L = \frac{n\pi^2 V_{Bat(Max)}}{(1+n\pi)\omega_p L_o}. \tag{16}$$

The total ripple current through the filter capacitor C_o is calculated considering M parallel rectifiers and taking into account the ripple cancellation effect due to the 180° phase displacement between the current through each inductor [31,32] in the current-doubler structure; thus,

$$\Delta i_C = \frac{n\pi^2 M V_{Bat(Max)}}{2(1+n\pi)\omega_p L_o} \tag{17}$$

The output voltage ripple is

$$\Delta v_{Bat} = \frac{n\pi^3 M V_{Bat(Max)}}{16(1+n\pi)\omega_p^2 L_o C_o} \tag{18}$$

From (18), the ripple of the charging current is a function of the switching frequency, output filter components, and battery parameters.

$$\Delta i_{Bat} = \frac{n\pi^3 M V_{Bat(Max)}}{16(1+n\pi) r_{But} \omega_p^2 L_v C_v} \tag{19}$$

The limitation of the output current ripple, Δi_{Bat}, is mandatory in order to avoid the battery degradation [12].

Reflected Impedance on the Primary Side of the Transformer

Since the output filter removes the high-frequency ripple, the low ripple approximation [29] is used to study the proposed rectifier in steady state. Considering the total current in the primary side and using the first harmonic of the square waveform, the relationship between the AC and DC currents is given in (20).

$$\hat{I}_{ac} = \frac{2 I_{Bat}}{n\pi} \tag{20}$$

where $\hat{I}_{ac}$ is the amplitude of the transformer's primary current. From (20) and (6), the charging current is obtained as a function of the angles $\Psi_0, \Psi_1, \ldots, \Psi_{N-1}$,

$$I_{Bat} = \frac{nV_{dc}}{Z_p}\sqrt{\left(\sum_{m=1}^{N}\sin\Psi_{m-1}\right)^2 + \left(\sum_{m=1}^{N}\cos\Psi_{m-1}\right)^2} \qquad (21)$$

The normalized amplitude of the charging current, I_{Bat}, is depicted in Figure 9 as a function of the control angle, Ψ, and considering the modulation pattern where all phases are evenly shifted.

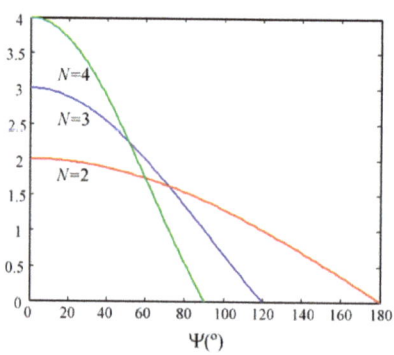

Figure 9. Amplitude of the normalized charging current, I_{Bat}, as a function of the control angle, Ψ, for N = 2, 3, and 4. All phases are evenly shifted, Ψ_0 = 0°, Ψ_1 = Ψ, Ψ_2 = 2Ψ, ..., Ψ_{N-1} = (N−1)Ψ.

Working with (21), the maximum charging current is achieved at $\Psi_0 = \Psi_1 = \Psi_{N-1} = 0°$ and is given by

$$I_{Bat(Max)} = \frac{nV_{dc}}{Z_p} \cdot N \qquad (22)$$

From (22), it can be observed that the output current capability of the multiphase converter is enhanced by increasing the number, N, of paralleled phases. An accurate acquisition of the modulation angle, covering the whole range over the entire battery charging process, facilitates the computation of ampere-hours in order to calculate the supplied capacity.

The amplitude of the voltage in the primary side of the transformer is obtained from the power balance in the windings. Assuming a lossless transformer,

$$P_{ac} = P_{Bat} = \frac{\hat{I}_{ac}\hat{V}_{ac}}{2} = I_{Bat}V_{Bat} \qquad (23)$$

and substituting (20) into (23),

$$\hat{V}_{ac} = n\pi V_{Bat} = n\pi r_{Bat} \cdot I_{Bat} + n\pi \cdot V_{Bat}. \qquad (24)$$

From (20) and (24), the battery is modeled from the AC side by

$$\hat{V}_{ac} = \frac{n^2\pi^2}{2} r_{Bat} \cdot \hat{I}_{ac} + n\pi \cdot V_{Bat}. \qquad (25)$$

The reflected impedance of the current multiplier and load, R_{ac}, into the AC side of the converter defines important characteristics of the resonant inverter, such as the switching mode of the transistors, the distortion of the waveforms, and the efficiency [11]. From (25), the rectifier stage is reflected into the AC side as the equivalent resistance R_{ac} in (26),

$$R_{ac} = \frac{\pi^2}{2} n^2 R_{Bat} = \frac{\pi^2}{2} n^2 \left(r_{Bat} + \frac{V_{Bat}}{I_{Bat}} \right) \quad (26)$$

Assuming an ideal transformer, where the leakage inductance reflected in the secondary side is $L_{ks} = 0$, the maximum voltage across the diodes is $V_B = -n\pi V_{Bat}$. However, in practice, L_{ks} is in series with the junction capacitance of the reverse-biased diode, C_j, causing a high-frequency oscillation or ringing. The selection of the Schottky devices takes into account the minimization of this effect.

5. Efficiency of the Multiphase LC_pC_s Resonant Converter

The overall efficiency of the converter is calculated by

$$\eta = \eta_I \cdot \eta_R, \quad (27)$$

where η_I is the efficiency of the resonant inverter stage and η_R is the efficiency of the output current multiplier stage.

5.1. Efficiency of the Inverter Stage

Taking into account the ZVS mode operation of the converter, the switching loss is considered negligible in comparison to the conduction loss. The efficiency of the resonant inverter stage, η_I, considering the conduction loss only [16] is

$$\eta_I = \frac{1}{1 + \frac{r}{R_{ac}} \cdot \frac{\sum_{k=1}^{N} \hat{I}_i^2}{\hat{I}_{ac}^2}} \quad (28)$$

where $\hat{I}_i$ is the amplitude of each resonant current given in (7). The resistance r represents the rds_{on} of the transistors as well as the ESR of the inductors. The highest efficiency, $\eta_{I(Max)}$, is achieved with $\Psi_0 = \Psi_1 = \ldots = \Psi_{N-1} = 0°$. Upon substitution of $\Psi_0 = \Psi_1 = \ldots = \Psi_{N-1} = 0°$ in (28) and under the assumption that C_s is calculated according to (9), the maximum efficiency as a function of the ratio r/R_{ac}, the nominal value of the quality factor, Q_{pN}, and the number of phases, N, is obtained.

$$\eta_{I(Max)} = \frac{1}{1 + \frac{r}{NR_{ac}} \cdot \left[1 + Q_{pN}^2 \right]} \quad (29)$$

From (28), it is observed that $\eta_{I(Max)}$ is improved by increasing R_{ac}. The straightforward way to increase R_{ac} is through the larger transformer turns ratio, n. However, it should be considered that Q_{pN} increases with n, according to (12), which could jeopardize the ZVS mode of the converter transistors. Taking into account the tight correlation among, N, Q_{pN}, n, and r/R_{ac}, the design process oriented to find a suitable value of these parameters involves iterative cycles. Upon the substitution of (12) and (20) into (29), $\eta_{I(Max)}$ is obtained as a function of the converter parameters,

$$\eta_{I(Max)} = \frac{1}{1 + \frac{\pi^2 r I_{Bat(Max)} V_{Bat(Max)}}{2 N V_{dc}^2} + \frac{2 r I_{Bat(Max)}}{n^2 \pi^2 N V_{Bat(Max)}}} \approx \frac{1}{1 + \frac{2 r I_{Bat(Max)}}{n^2 \pi^2 N V_{Bat(Max)}}} \quad (30)$$

The maximum efficiency of the resonant inverter stage, $\eta_{I(Max)}$, improves, approaching one asymptotically as the number of phases, N, increases.

5.2. Efficiency of the Output Current Multiplier

Limiting the current level through the output rectifier stage is a major design challenge oriented to reduce the conduction loss. The proposed M-windings output current multiplier lowers the amplitude of the current through diodes by a factor M and the average current through filters inductors by a factor $2M$. An expression for the rectifier efficiency, η_R, only including the conduction loss, is obtained from the analysis of the current paths shown in Figure 7. Considering a lossless transformer, the total power, P_T, in the secondary side of the current multiplier is

$$P_T = P_{Bat} + M\left(\frac{V_D I_{Bat}}{M} + \frac{I_{Bat}^2 r_D}{M^2} + \frac{I_{Bat}^2 r_{LF}}{4M^2}\right), \quad (31)$$

where P_{Bat} is the output power, $P_{Bat} = V_{Bat} \cdot I_{Bat}$, V_D and r_D are the voltage and dynamic resistance of the linear model of the diode, and r_{LF} is the ESR of the filter inductor L_o. The efficiency, η_R, is calculated with $\eta_R = P_{Bat}/P_T$,

$$\eta_R = \frac{1}{1 + \frac{V_D}{V_{Bat}} + \frac{(\frac{r_D}{M} + \frac{r_{LF}}{2M})I_{Bat}}{V_{Bat}}}. \quad (32)$$

The efficiency of the output current multiplier, η_R, is improved by increasing the number of secondary windings, M. The theoretical limit $\eta_{R(Max)}$ of η_R is obtained letting $M \to \infty$,

$$\eta_{R(Max)} = \frac{1}{1 + \frac{V_D}{V_{Bat(Max)}}}. \quad (33)$$

From (32) and (33), it can be observed that the ratio $V_D/V_{Bat(Max)}$ should be minimized, which confirms the benefit of using Schottky diodes or sync rectifiers to improve the efficiency of the rectifier stage.

5.3. Optimum N and M of Parallelized Stages

The expressions (30) to (33) are used as a criterion to define the appropriate number of phases, N, of the resonant inverter stage as well as the number of secondary windings, M, of the output current multiplier. The maximum efficiency of the inverter section, $\eta_{I(Max)}$, and the efficiency of the current multiplier, η_R, are depicted in Figure 10 as a function of N and M.

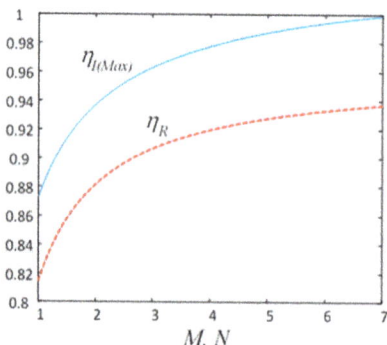

Figure 10. Maximum efficiency of the resonant inverter stage, $\eta_{I(Max)}$, and efficiency of the output current multiplier, η_R, as a function of the number of phases, N, and secondary windings M. The following typical values are assumed: $n = 2$, $r = 1.7\ \Omega$, $I_{Bat(Max)} = 20$ A, $2rI_{Bat(Max)}/\pi^2 V_{Bat(Max)} = 0.5$, $V_D/V_{Bat(Max)} = 0.05$ and $r_D = r_{LF} = 0.1$.

From a practical point of view, the asymptotic variation of η_R and $\eta_{I(Max)}$, shown in Figure 10, limits the maximum values of M and N. The criterion for choosing the suitable values of M and N is a tradeoff between the increment of the efficiency and the circuit complexity. It is assumed that if the increment of efficiency achieved is barely 1%, a higher number of secondary windings M or phases, N, is not justified.

6. Design of the Multiphase LC_pC_s Resonant Converter

(1) The maximum battery voltage is set at $V_{Bat(Max)} = 53.5$ V, which is below the overvoltage protection limit (54.7 V) defined by the BMS. The output current capability of the circuit is set to $I_{Bat} = 20$ A in order to shortening the charging time. The equivalent impedance of the battery is $R_{Bat} = 2.67$ Ω. The peak power that must be supplied by the charger is $P_{Bat} = 1.07$ kW. The converter supply voltage is $V_{dc} = 400$ V, which is the output voltage of a previous front-end PFC stage. The switching frequency is set at $\omega_p = 2\pi(125 \text{ kHz})$.

(2) The drive signals of the transistors are obtained from an integrated circuit IR2111 with a dead time, $t_d = 650$ ns. From (10), the minimum value of the power factor angle for each class D section is $\phi_{zvs} = 29.25°$. Using the design constrain $\varphi_i = 2\phi_{zvs} = 58°$ from (11), the nominal value of the quality factor is obtained, $Q_{pN} = 0.624$. The transformer turns ratio, n, is calculated from (14), approximating to the nearest entire value, $n = 1$.

(3) The number of phases, N, is calculated taking into account that transistors are low-cost CoolMOS™ SPA11N60C3 (Infineon, Neubiberg, Germany) with $rds_{(on)} = 0.38$ Ω. Considering the equivalent series resistant (ESR) of the resonant inductors and tracks of the printed circuit board (PCB), a worst case of $r = 1$ Ω is assumed. Upon substitution in (30), the pair $n = 1$ and $N = 4$ yields $\eta_{I(Max)} = 0.98$. This value of efficiency means 21 W power loss in the resonant inverter stage at full load conditions.

(4) The expected efficiency of the rectifier stage is calculated using the conduction loss model of the Schottky diode STPS30M60S (STMicroelectronics, Geneva, Switzerland) from ST with $V_D = 0.395$ V and $r_D = 0.0047$ Ω. The filter inductors are Vishay IHLP–8787MZ (Vishay Intertechnology, Malvern, USA) with $L_o = 75$ μH and $r_{LF} = 30$ mΩ at 25° C. Taking into account the temperature effect, the value $r_{LF} = 90$ mΩ is assumed. Upon substitution of V_D, r_D, $V_{Bat(Max)}$, r_{LF}, and $I_{Bat} = 20$ A in (32), the value $M = 1$ yields an efficiency of the rectifier stage at maximum load, $\eta_R = 0.97$. This value of efficiency means 32 W power loss in the current-doubler rectifier at full load. In this way, the configuration of a four-phases ($N = 4$) resonant inverter with a single ($M = 1$) current-doubler rectifier as output stage achieves an overall efficiency at full load equal to $\eta = \eta_I \cdot \eta_R = 0.95$.

(5) From (16), the amplitude of the current ripple in each inductor is $\Delta i_L = 2.16$ A. The r_{Bat} is estimated at 40 mΩ. The output capacitor, C_o, is calculated to achieve a maximum current ripple equal to 0.1% of the charging current, $\Delta i_{Bat} = 20$ mA. From (19), $C_o = 680$ μF.

(6) The characteristic impedance is obtained from (22), $Z_p = 80$ Ω. In Table 1, the reactive components are $L = Z_p/\omega_p = 100$ μH and $C_p = 4/\omega_p Z_p = 64$ nF.

(7) The transformer has been built with an ETD49 core of material N87. The primary and secondary are 16 single-layer turns of 40 strands of litz wire. The resulting magnetizing inductance is $L_M = 800$ μH and the leakage inductance from the primary and secondary sides are $L_{kp} = L_{ks} = 1.4$ μH. The total leakage inductance is $L_k = L_{kp} + n^2 \cdot L_{ks} = 2.8$ μH.

(8) Once L_k is known, the series capacitor C_s is calculated with (9) to cancel out the effect of L_k, $C_s = 571$ nF.

7. Control Circuit and Battery Modeling

During the CV stage, the charging current must be regulated to avoid the voltage of the battery exceeding $V_{Bat(Max)}$. The current is modulated through the phase-angles Ψ_0, Ψ_1, and Ψ_{N-1}, while keeping the switching frequency constant. Different patterns are possible for adjusting Ψ_1, Ψ_2, and Ψ_{N-1}. For any value of N, the full control of the charging current is achieved if the phase shift is evenly distributed among all N phases, e.g., $\Psi_0 = 0°$, $\Psi_1 = \Psi$, $\Psi_2 = 2\Psi \dots \Psi_{N-1} = (N-1)\Psi$. In this case, the minimum current $I_{Bat} = 0$ A is achieved at $\Psi = 360°/N$. This pattern requires N control signals. For this design, where $N = 4$, the control angles are adjusted as follows: $\Psi_0 = \Psi_1 = 0°$ and $\Psi_2 = \Psi_3 = \Psi$. For this approach, the minimum $I_{Bat} = 0$ A is achieved at $\Psi = 180°$, and only two control signals are required, which implies a simplification of the control circuit.

Once the converter is designed, the battery-charger system is completed with a control loop to limit the output voltage of the charger to the maximum value recommended for the battery. The action of the control loop transforms the circuit's open-loop current source behavior into a voltage source. A type I error amplifier is enough for this action. The scheme of the charger-battery system, modeled in Simulink, is shown in Figure 11. In the voltage mode, the battery imposes the dynamic response of the converter-battery system [27].

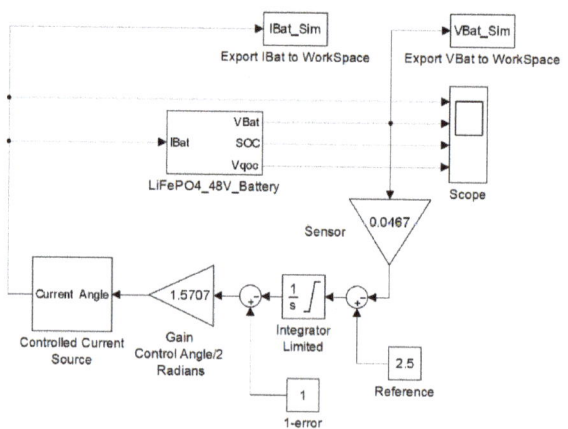

Figure 11. Control loop for limiting the maximum battery voltage, $V_{Bat(Max)}$.

The Simulink® model of the battery [34,35] is shown in Figure 12. The look-up tables include the quasi-open circuit voltage of a basic cell for the charge and discharge trajectories as a function of the *SOC*.

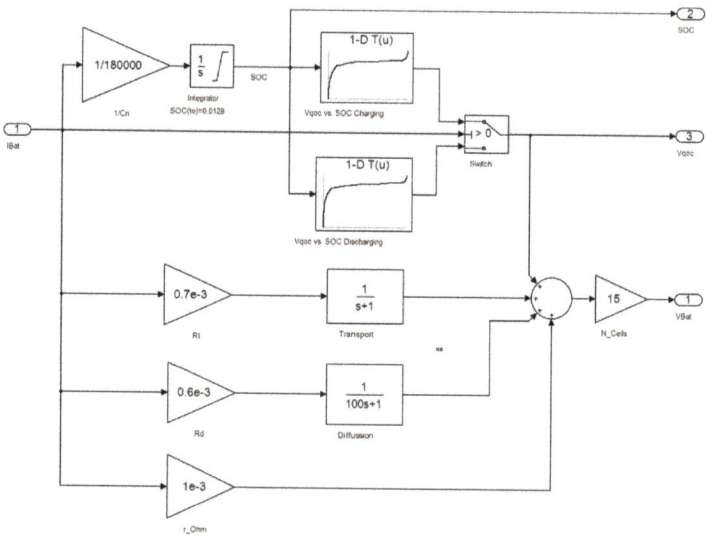

Figure 12. Simulink® model of the battery.

The different parameters of the model can be tuned using curve fitting. The data used as reference for adjusting the model were obtained from the experimental characterization of the battery charging at 25 A, which has been shown in Figure 1. The time constant for charge transportation and diffusion phenomena are 1 s and 100 s, respectively. The impedance for the charge transport is R_t = 0.7 mΩ and the capacitance is C_t = 1428 F. The impedance of the diffusion is R_d = 0.6 mΩ and the corresponding capacitance is C_d = 166,000 F. The impedance due to electrodes and electric connections is R_Ω = 1 mΩ. The impedance of the battery pack is obtained from (4), r_{bat} = 34.5 mΩ. This value of r_{Bat} includes the impedance of connectors and cables, which is used to conform to the battery by the series connection of the 15 cells.

The variation of the battery voltage, obtained from the simulation of the system in Figure 11, is shown in Figure 13. It can be observed that simulation and experimental results are in good agreement for the three charging profiles in Figure 2 that were evaluated experimentally.

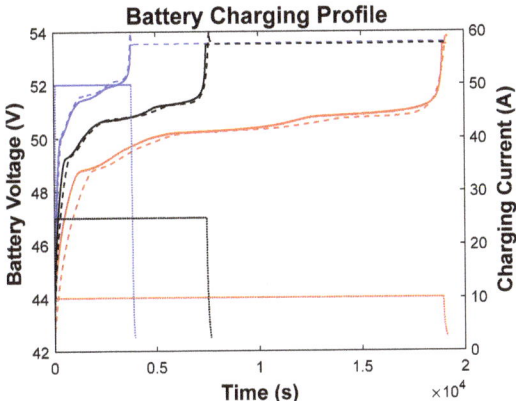

Figure 13. Charging profiles at 10, 25, and 50 A. Solid lines: Experimental battery voltage. Dashed lines: Simulation result. Dot lines: Charging current.

8. Results of the Experimental Prototype

An experimental prototype, shown in Figure 14, has been built to validate the theoretical proposal.

Figure 14. Details of the experimental prototype of the charger. (**Left**) Four-phase resonant inverter stage. (**Right**) Current-doubler rectifier.

When connecting the battery to the charger, an initial frequency sweep is programmed to ensuring the gradual growth of the charging current to prevent the occurrence of an overvoltage across the discharged battery. The experimental waveforms in different circuit sections are shown in Figures 15–17. In order to demonstrate the charger performance at different operation points, the waveforms for full load and 70% of full load operation are shown. In Figure 15, it is observed that the resonant current has a phase lag with respect to the input voltage. At full load condition, $\varphi_{i1,2} = \varphi_{i3,4} = 54°$, which is in good agreement with the theoretical value, and at 70% of the full load condition, $\varphi_{i1,2} = 54°$, $\varphi_{i3,4} = 72°$. The ZVS mode operation was verified for all phases of the resonant inverter section.

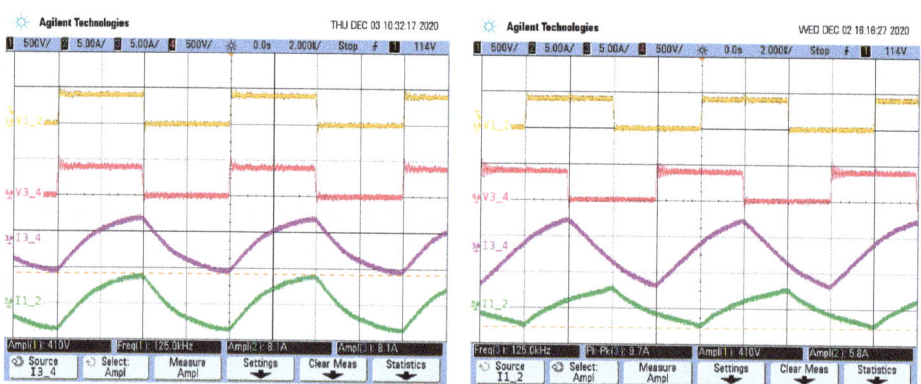

Figure 15. From top to bottom: Midpoint voltages of phases 1 and 2, $v_{1,2}$. Midpoint voltages of phases 3 and 4, $v_{3,4}$. Resonant current of phases 1 and 2, $i_{1,2}$. Resonant current of phases 3 and 4, $i_{3,4}$. (**Left**) Full load condition ($\Psi_0 = \Psi_1 = \Psi_2 = \Psi_3 = 0°$). (**Right**) 70% of full load condition ($\Psi_0 = \Psi_1 = 0°, \Psi_2 = \Psi_3 = 90°$).

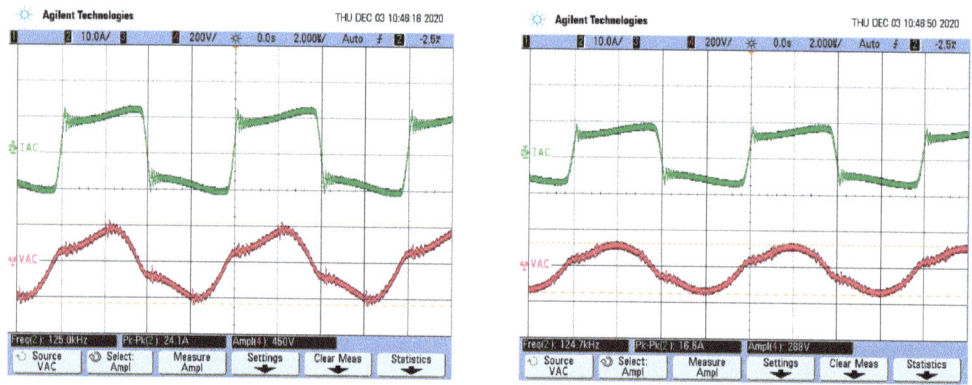

Figure 16. From top to bottom: Output current through the primary side of the transformer, i_{ac}. Output voltage applied to the primary side of the transformer, v_{ac}. (**Left**) Full load condition ($\Psi_0 = \Psi_1 = \Psi_2 = \Psi_3 = 0°$). (**Right**) 70% of full load condition ($\Psi_0 = \Psi_1 = 0°, \Psi_2 = \Psi_3 = 90°$).

In Figure 16, the current and voltage at the primary side of the transformer are shown. The amplitude of the current square waveform is half (10 A) of the battery charging current. In Figure 17, the charging current at full load (20 A) and at 70% of full load are shown. The results are in good agreement with the theoretical value according to the control angle Ψ. It can be observed that the charging current ripple is negligible as it is required for this application. The experimental efficiency of the prototype measured at the point of maximum load (I_{Bat} = 20 A, P_{Bat} = 1.07 kW) was η = 91.3%. The efficiency at 70% and 50% of the full load was η = 90.2% and η = 88%, respectively. The experimental efficiency is slightly lower than the theoretical due to the switching losses, the power dissipation at the transistors drive circuit, and the auxiliary power supply loss.

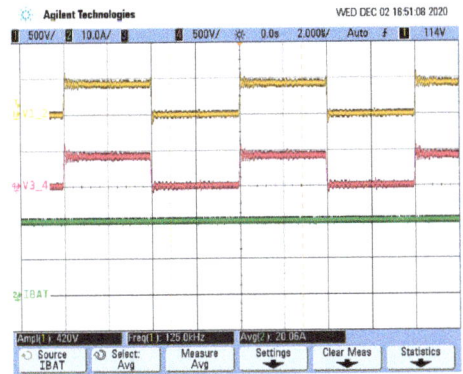
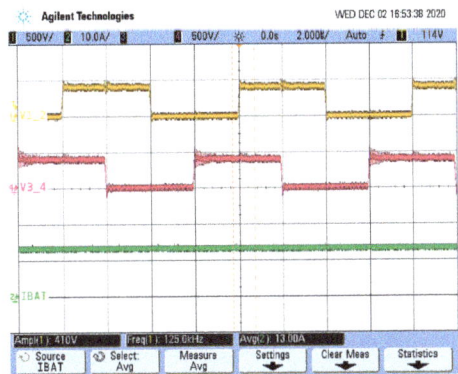

Figure 17. From top to bottom: Midpoint voltages of phases 1 and 2, $v_{1,2}$. Midpoint voltages of phases 3 and 4, $v_{3,4}$. Charging current i_{Bat}. (**Left**) Full load condition ($\Psi_0 = \Psi_1 = \Psi_2 = \Psi_3 = 0°$). (**Right**) 70% of full load condition ($\Psi_0 = \Psi_1 = 0°$, $\Psi_2 = \Psi_3 = 90°$).

9. Discussion

In this work, the general design method of the proposed charger has been explained, but the particular configuration of the final solution depends on the chosen technology. One key decision is the most suitable value of the dc-link voltage. The solution for an dc-link voltage V_{dc} = 400 V, which was obtained from a single-phase power factor corrector (PFC) based on a Boost Converter, and using the CoolMOS^{TM} SPA11N60C3 MOSFET transistor and the STPS30M60S Schottky diode has been fully developed. As alternative, a solution with V_{dc} = 800 V, obtained from a three-phase PFC and using silicon carbide (SiC) components is also assessed. For this case, the third-generation C3M0065100K MOSFET transistor (Wolfspeed, Research Triangle Park, USA) with the CGD15SG00D2 driver (Wolfspeed, Research Triangle Park, USA) is used in the inverter section. As the voltage, current, and power at the circuit output are the same, the silicon (Si) Schottky diode STPS30M60S is used in both cases. For a better comparison, both designs are summarized in Table 2.

Table 2. Designs comparison.

V_{dc}	n	N	M	Z_p	Q_{pN}	η	$I_{Bat(Max)}$	$P_{Bat(Max)}$
400 V	1	4	1	80 Ω	0.624	0.95	20 A	1.07 kW
800 V	2	2	1	160 Ω	0.624	0.966	20 A	1.07 kW

As it can be seen in Table 2, both designs achieve a similar theoretical efficiency, but the SiC technology uses only two phases for the resonant inverter stage.

Considerations about the Solution Cost

SiC technology for power devices is becoming more competitive in technical performance and cost. Important advances have been reported in terms of increasing the wafer diameter and minimization of the defect density [36], which contribute to lowering the cost of the devices, so it is worth comparing the cost of the proposed alternatives. Focusing on the inverter section of the described designs, i.e., V_{dc} = 400 V for the four-phase Si inverter and V_{dc} = 800 V for the two-phase SiC inverter, the cost assessment reveals that at present, the solution based on SiC components is more expensive despite requiring fewer transistors. The cost of the third-generation SiC MOSFET C3M0065100K is five times (5 ×) that of the SPA11N60C3 Si MOSFET. On the other hand, in contrast to the simplicity of the half-bridge driver, based on the integrated circuit IR2111, the complexity and cost of the selected driver CGD15SG00D2 for SiC MOSFETS are also significantly higher [37]. In addition, the PFC section adds a cost difference in favor of the V_{dc} = 400 V four-phase Si design. For illustrating the analysis, in Tables 3 and 4, the cost of the SiC components and its Si counterparts are summarized [38]. Differences in the magnetic elements, capacitors, and control circuit have less impact on cost.

Table 3. SiC resonant inverter.

Component	Quantity	Cost (Retail Sale)
MOSFET C3M0065100K	4	40€
Driver CGD15SG00D2	4	200€

Table 4. Si resonant inverter.

Component	Quantity	Cost (Retail Sale)
MOSFET SPA11N60C3	8	16€
Driver IR2111	4	8€

Nowadays, for a given architecture, the use of SiC MOSFETs could be recommended if the maximum current, voltage, and temperature limits of the Si MOSFETs are compromised, e.g., for charging currents and powers higher than 50 A and 2.5 kW, respectively.

10. Conclusions

The general design procedure of a multiphase resonant converter for battery charger applications has been presented. Since the output current on the AC side is shared among N equal inverter sections, the circuit presents high output current capability using low-cost power MOSFETs, and the design of the resonant inductors is simplified. The proposed output rectifier is based on an M-winding current-doubler rectifier that also diminishes the conduction loss by using passive components. The efficiency curve of the proposed charger exhibits a wide flat zone, assuring a constant value of efficiency even at light load conditions. This feature is very interesting for the battery charger applications, taking into account that high efficiency is desirable along the whole charging process, despite the heavy load variation. The effect on the AC side of the leakage inductance of the transformer L_k is canceled out by the series capacitor C_s. The maximum charging current is limited by the circuit in an inherent manner, without the necessity of any control. However, the output voltage is limited to the maximum value recommended for the battery by a voltage control loop with a type I error amplifier. The control action is performed keeping constant the switching frequency by adjusting the control angle, Ψ, while maintaining the ZVS mode at any operation point. The general proposal has been validated by implementing an experimental prototype for charging a commercial 48 V LiFePO$_4$ battery with 50 Ah of capacity. The achieved efficiency of the $N = 4$ inverter with $V_{dc} = 400$ V using Si MOSFETs is similar to the predicted with an $N = 2$ inverter with $V_{dc} = 800$ V using SiC MOSFETs.

Author Contributions: The battery characterization and the experimental evaluation of the different charging profiles were carried out at the Batteries Laboratory of University of the Oviedo by J.C.V. and M.G. The proposed charger was developed at the Power Electronics Laboratory of the University of Cantabria by C.B., F.J.A., R.C. and F.J.D. Conceptualization, C.B. and J.C.V.; methodology, C.B. and J.C.V.; validation, C.B., J.C.V., R.C. and F.J.D.; formal analysis, C.B. and J.C.V.; investigation, C.B., J.C.V. and F.J.A.; writing—original draft preparation, C.B. and J.C.V.; writing—review and editing, F.J.A., M.G., R.C. and F.J.D.; project administration, F.J.A. and M.G.; funding acquisition, F.J.A. and M.G. All authors have read and agreed to the published version of the manuscript.

Funding: This work was funded by the Spanish Ministry of Science and the EU through the projects RTI2018-095138-B-C31: "Power Electronics for the Grid and Industry Applications", TEC2016-80700-R (AEI/FEDER/UE), PID2019-110955RB-I00, and by the Principality of Asturias via Project FC-IDI/2018/000226.

Conflicts of Interest: The authors declare no conflict of interest.

References

1. Khaligh, A.; Li, Z. Battery ultracapacitor, fuel cell, and hybrid energy storage systems for electric, hybrid electric, fuel cell, and plug-in hybrid electric vehicles: State of the art. *IEEE Trans. Veh. Technol.* **2010**, *59*, 2806–2814. [CrossRef]
2. Braun, P.; Cho, J.; Pikul, J.; King, W.; Zhang, H. High power rechargeable batteries. *Curr. Opin. Solid State Mater. Sci.* **2012**, *16*, 186–198. [CrossRef]
3. Waag, W.; Fleischer, C.; Sauer, D.U. Critical review of the methods for monitoring of lithium-ion batteries in electric and hybrid vehicles. *J. Power Sources* **2014**, *258*, 321–339. [CrossRef]
4. Wei, Z.; Zhao, J.; Xiong, R.; Dong, G.; Pou, J.; Tseng, K.J. Online Estimation of Power Capacity With Noise Effect Attenuation for Lithium-Ion Battery. *IEEE Trans. Ind. Electron.* **2019**, *66*, 5724–5735. [CrossRef]
5. Wei, Z.; Zhao, D.; He, H.; Cao, W.; Dong, G. noise-tolerant model parameterization method for lithium-ion battery management system. *Appl. Energy* **2020**, *268*, 114932. [CrossRef]
6. Ke, M.Y.; Chiu, Y.H.; Wu, C.Y. Battery Modelling and SOC Estimation of a LiFePO$_4$ Battery. In Proceedings of the 2016 International Symposium on Computer, Consumer and Control (IS3C), Xi'an, China, 4–6 July 2016.
7. Wang, A.; Jin, X.; Li, Y.; Li, N. LiFePO$_4$ battery modeling and SOC estimation algorithm. In Proceedings of the 2017 29th Chinese Control and Decision Conference (CCDC), Chongqing, China, 28–30 May 2017.
8. Kazimierczuk, M.K.; Czarkowski, D. *Resonant Power Converters*, 2nd ed.; John Wiley & Sons: Hoboken, NJ, USA, 2012.
9. Lin, C.H.; Wang, C.M.; Hung, M.H.; Li, M.H. Series-resonant battery charger with synchronous rectifiers for LiFePO$_4$ battery pack. In Proceedings of the 2011 6th IEEE Conference on Industrial Electronics and Applications, Beijing, China, 21–23 June 2011.
10. Karimi, S.; Tahami, F. A Comprehensive Time-domain-based Optimization of a High-Frequency LLC-based Li-ion Battery Charger. In Proceedings of the 2019 10th International Power Electronics, Drive Systems and Technologies Conference (PEDSTC), Shiraz, Iran, 12–14 February 2019.
11. Shafiei, N.; Ordonez, M.; Craciun, M.; Botting, C.; Edington, M. Burst Mode Elimination in High-Power LLC Resonant Battery Charger for Electric Vehicles. *IEEE Trans. Power Electron.* **2016**, *31*, 1173–1188. [CrossRef]
12. Park, S.M.; Kim, D.H.; Joo, D.M.; Kim, M.J.; Lee, B.K. Design of Output Filter in LLC Resonant Converters for Ripple Current Reduction in Battery Charging Applications. In Proceedings of the 9th International Conference on Power Electronics-ECCE Asia, Seoul, Korea, 1–5 June 2015.
13. Liu, X.; Baguley, C.A.; Madawala, U.K.; Thrimawithana, D.J. A Compact Power Converter for High Current and Low Voltage Applications. In Proceedings of the 39th Annual Conference of the IEEE Industrial Electronics Society, Vienna, Austria, 10–13 November 2013; pp. 140–144.
14. Yilmaz, M.; Krein, P.T. Review of battery charger topologies, charging power levels, and infrastructure for plug-in electric and hybrid vehicles. *IEEE Trans. Power Electron.* **2013**, *28*, 2151–2169. [CrossRef]
15. Ashok, B. Wide-Bandgap-Based Power Devices. *IEEE Power Electron. Mag.* **2015**, *2*, 42–47.
16. Branas, C.; Azcondo, F.J.; Casanueva, R. A Generalize Study of Multiphase Parallel Resonant Inverters for High-Power Applications. *IEEE Trans. Circuits Syst.* **2008**, *55*, 2128–2138. [CrossRef]
17. Narada NPFC Series. Operation Manual. Available online: https://mpinarada.com/wp-content/uploads/2019/01/OM-Narada-NPFC-Series-Li-Ion-0423-V8.pdf (accessed on 22 January 2021).
18. Keil, P.; Jossen, A. Charging protocols for lithium-ion batteries and their impact on cycle life—An experimental study with different 18650 high-power cells. *J. Energy Storage* **2016**, *6*, 125–141. [CrossRef]
19. Wu, J.; Wei, Z.; Liu, K.; Quan, Z.; Li, Y. Battery-Involved Energy Management for Hybrid Electric Bus Based on Expert-Assistance Deep Deterministic Policy Gradient Algorithm. *IEEE Trans. Veh. Technol.* **2020**, *69*, 12786–12796. [CrossRef]
20. Wu, J.; Wei, Z.; Li, W.; Wang, Y.; Li, Y.; Sauer, D. Battery Thermal- and Health-Constrained Energy Management for Hybrid Electric Bus based on Soft Actor-Critic DRL Algorithm. *IEEE Trans. Ind. Inform.* **2020**. [CrossRef]
21. Notten, P.H.; het Veld, J.O.; Van Beek, J.R.G. Boostcharging Li-ion batteries: A challenging new charging concept. *J. Power Sources* **2005**, *145*, 89–94. [CrossRef]
22. Perez, H.E.; Dey, S.; Hu, X.; Moura, S.J. Optimal Charging of Li-Ion Batteries via a Single Particle Model with Electrolyte and Thermal Dynamics. *J. Electrochem. Soc.* **2017**, *164*, 1–10. [CrossRef]
23. Anseán, D.; González, M.; Viera, J.C.; García, V.M.; Blanco, C.; Valledor, M. Fast charging technique for high power lithium iron phosphate batteries: A cycle life analysis. *J. Power Sources* **2013**, *239*, 9–15. [CrossRef]
24. Anseán, D.; Dubarry, M.; Devie, A.; Liaw, B.Y.; García, V.M.; Viera, J.C.; González, M. Fast charging technique for high power LiFePO$_4$ batteries: A mechanistic analysis of aging. *J. Power Sources* **2016**, *321*, 201–209. [CrossRef]
25. Chen, M.; Rincon-Mora, G.A. Accurate Electrical Battery Model Capable of Predicting Runtime and I-V Performance. *IEEE Trans. Energy Convers.* **2006**, *21*, 504–511. [CrossRef]

26. Truchot, C.; Dubarry, M.; Liaw, B.Y. State-of-charge estimation and uncertainty for lithium-ion battery strings. *Appl. Energy* **2014**, *119*, 218–227. [CrossRef]
27. Jossen, A. Fundamentals of battery dynamics. *J. Power Sources* **2006**, *154*, 530–538. [CrossRef]
28. Bojarski, M.; Asa, E.; Colak, K.; Czarkowski, D. A 25 kW Industrial Prototype Wireless Electric Vehicle Charger. In Proceedings of the 2016 IEEE Applied Power Electronics Conference and Exposition (APEC), Long Beach, CA, USA, 20–24 March 2016; pp. 1756–1761.
29. Erickson, R.W.; Maksimovic, D. *Fundamentals of Power Electronics*, 2nd ed.; Springer: New York, NY, USA, 2001.
30. Zhang, J.; Hurley, W.G.; Wölfle, W.H. Gapped Transformer Design Methodology and Implementation for LLC Resonant Converters. *IEEE Trans. Ind. Appl.* **2016**, *52*, 342–350. [CrossRef]
31. Huber, L.; Jovanovic, M.H. Forward-flyback Converter with Current-Doubler Rectifier: Analysis, Design and Evaluation Results. *IEEE Trans. Power Electron.* **1999**, *14*, 184–192. [CrossRef]
32. Alou, P.; Oliver, J.A.; García, O.; Prieto, R.; Cobos, J.A. Comparison of Current Doubler Rectifier and Center Tapped Rectifier for Low Voltage Applications. In Proceedings of the Twenty-First Annual IEEE Applied Power Electronics Conference and Exposition, (APEC'06), Dallas, TX, USA, 19–23 March 2006; pp. 744–750.
33. Lopez, V.M.; Navarro-Crespin, A.; Schnell, R.W.; Branas, C.; Azcondo, F.J.; Zane, R. Current Phase Surveillance in Resonant Converters for Electric Discharge Applications to Assure Operation in Zero-Voltage-Switching Mode. *IEEE Trans. Power Electron.* **2012**, *27*, 2925–2935. [CrossRef]
34. Plett, G. Extended Kalman Filtering for Battery Management System of LiPB-Based HEV Battery Packs—Part 1: Background. *J. Power Sources* **2004**, *134*, 252–261. [CrossRef]
35. Plett, G. Extended Kalman Filtering for Battery Management Systems of LiPb-Based HEV Battery Packs—Part 2: Modeling and Identification. *J. Power Sources* **2004**, *134*, 262–276. [CrossRef]
36. Loboda, M.J.; Chung, G.; Carlson, E.; Drachev, R.; Hansen, D.; Sanchez, E.; Wan, J.; Zhang, J. Advances in SiC Substrates for Power and Energy Applications. In Proceedings of the GOMACTECH Conference, Palm Springs, CA, USA, 16–19 May 2011.
37. Hazra, S.; Vechalapu, K.; Madhusoodhanan, S.; Bhattacharya, S.; Hatua, K. Gate Driver Design Considerations for SiliconCarbide MOSFETs Including For Series Connected Devices. In Proceedings of the 2017 IEEE Energy Conversion Congress and Exposition (ECCE), Cincinnati, OH, USA, 1–5 October 2017.
38. Nielsen, R.Ø.; Török, L.; Munk-Nielsen, S.; Blaabjerg, F. Efficiency and Cost Comparison of Si IGBT and SiC JFET Isolated DC/DC Converters. In Proceedings of the IECON 2013—39th Annual Conference of the IEEE Industrial Electronics Society, Vienna, Austria, 10–13 November 2013.

Article

A Novel High-Efficiency Double-Input Bidirectional DC/DC Converter for Battery Cell-Voltage Equalizer with Flyback Transformer

Fengdong Shi [1,2] and Dawei Song [1,2,*]

1. School of Electrical Engineering and Automation, Tiangong University, Tianjin 300387, China; shifengdong@tjpu.edu.cn
2. Center for Engineering Internship and Training, Tiangong University, Tianjin 300387, China
* Correspondence: songdawei@tjpu.edu.cn

Received: 30 October 2019; Accepted: 27 November 2019; Published: 29 November 2019

Abstract: Large-scale battery cells are connected in series, which inevitably leads to a phenomenon that the cell voltage is unbalanced. With a conventional equalizer, it is challenging to maintain excellent characteristics in terms of its size, design cost, and equalization efficiency. In order to improve the defects in the above equalization circuit, a novel voltage equalization circuit is designed, which can work in two modes. A bidirectional direct current–direct current (DC–DC) equalization structure is adopted, which can quickly equalize two high or low-power batteries without using an external energy buffer. In order to verify the effectiveness of the proposed circuit, a 12-cell battery 2800-MAh battery string was applied for experimental verification. Computer monitoring (LabVIEW) was adopted in the whole system to intelligently adjust the energy imbalance of the battery pack. The experimental results showed excellent overall performance in terms of equalization was achieved through the newly proposed method. That is, the circuit equalization speed, design cost, and volume have a good balance performance.

Keywords: battery equalization; flyback transformer; topology

1. Introduction

Lithium-ion batteries have been widely used in the field of electric vehicles due to their high charging–discharging times and high energy efficiency. Since the battery pack in a pure electric vehicle is connected in series and in parallel through a large number of battery cells, some of the battery cells may be unbalanced during the continuous charging and discharging process of the battery. The overall performance of the battery pack will be restricted by the battery cells with the lowest battery capacity, which will seriously affect the service life of the battery pack and reduce the battery life of the electric vehicle [1,2]. Battery equalization technology can suppress imbalance in the battery pack, leading to improved work consistency. The battery management system is able to operate well, which can ensure safe driving of new energy vehicles, and the working principle is shown in Figure 1. Excess energy (Converted Energy in the legend) can be transferred from cell 1 to cell 2 through the energy converter. A relatively balanced state is achieved between the two cells, and so on, resulting in consistent energy obtained by the entire battery pack.

In recent years, plenty of battery pack equalization methods have been proposed [3–6]. The equalization method has been divided into different categories according to various criteria. The energy dissipation has two types [7,8]: switch shunt and fixed shunt. This equalization circuit with a simple structure was easy to control. However, the converted heat will result in increased temperature in the entire battery pack, and the system will require extra energy. Compared with the energy dissipation type, a peripheral energy conversion circuit was applied to balance the voltage of the battery through the non-energy dissipation equalization method. Although the circuit is more complex, it is more efficient and safe according to the different means of energy transmission. These methods are further subdivided into four types: cell-to-cell equalization methods, cell-to-pack equalization methods, pack-to-cell equalization methods, and cell-to-pack to cell equalization methods, as shown in Figure 2.

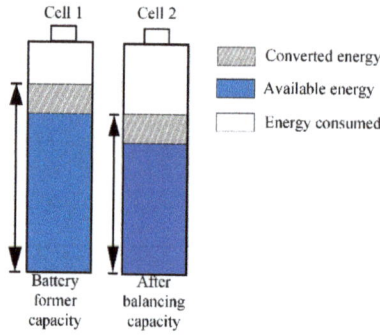

Figure 1. Battery balance diagram.

In the above equalization method, the monomer-to-monomer equalization method and the monomer-to-battery equalization method can effectively prevent overcharging and over-discharging of the battery. Even if the energy overlap and high-pressure stress in the equilibrium process are considered, the equilibrium efficiency cannot be guaranteed. In contrast, cell-to-cell equalization methods with short and efficient transmission paths are ideal choices. These methods are further subdivided into three types: transmission equalization methods, a non-isolated direct current–direct current (DC–DC) equalization method [9], and an isolated DC–DC equalization method [10]. One of the most conventionally used topologies of the transmission equalization methods is the bidirectional buck-boost converter [11]. Every two adjacent cells have a common buck-boost converter to achieve energy transfer. However, when the positions of the unbalanced batteries are not adjacent, the path of energy transfer becomes long. In order to compensate for the shortcomings of such equalization methods, Li, Y proposed a structure of an equalization method [12].

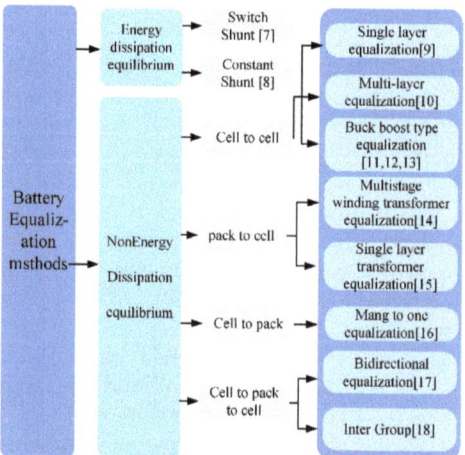

Figure 2. Battery balance-type diagram.

This circuit transfers energy from the most charged unit to the least charged unit to have higher equalization efficiency, but at the expense of lower equalization speed. Another direct cell-to-cell equalization method which uses an inductor as the energy storage component has fast equalization. However, each battery has two directional transmission channels, which require a large number of switches and diodes. Similarly, the direct inter-cell equalization method using a transformer ensures a fast equalization speed [13], but the use of a large number of transformers is costly. Obviously, the aforementioned direct cell-to-cell equalization method cannot achieve fast equalization speed and high efficiency, and relatively low cost.

In order to improve the deficiencies of these defects, a dual-winding battery equalizer with energy bidirectional flow was proposed. The advantage is that the cells with low energy are supplemented, and the cells with high electric quantity are effectively weakened. The principle of operation is an equalization structure using bidirectional forward and flyback equalization. Any two high-energy cells or low-energy cell in the battery pack was quickly equalized without external energy buffer. To realize the effective balance of the whole battery pack.

2. Proposed Equalization Topology

2.1. Proposed Equalizer

The proposed voltage equalizer is composed of a left and right sides interleaved switching network and an isolated double winding DC–DC converter. Figure 3 shows an equalization structure based on a flyback coaxial multi-winding output, which is mainly composed of a flyback transformer T_1 and a battery pack B_1-B_n circuit. The battery pack as a whole and single cells form a loop through the output winding of each channel. Energy is transmitted to the battery cell from one direction.

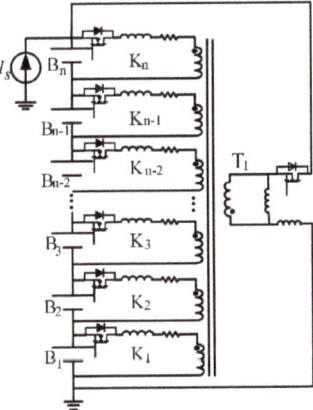

Figure 3. Coaxial multi-winding equalizer.

The equilibrium method proposed in this paper could be extended from the original single-to-single equilibrium to the multi-to-two equilibrium structure. The staggered switching network is combined with the two operation modes. Bidirectional equalization is realized in the energy-transfer process as shown in Figure 4. The left and right sides are staggered control switches by the application of a two-winding flyback transformer for output isolation. The bidirectional equalization equivalent circuit is shown in Figure 5.

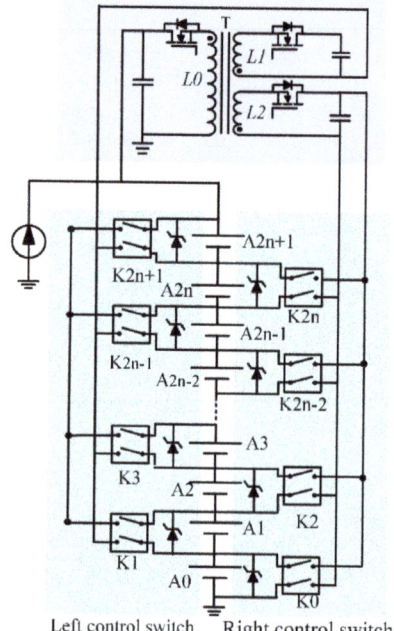

Figure 4. New voltage equalizer.

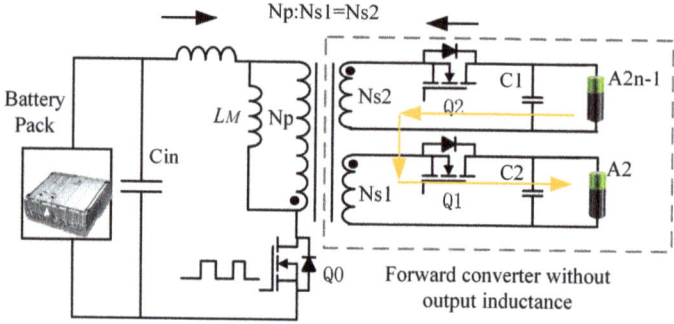

Figure 5. Equalization equivalent circuit.

Each cell is connected with a control switch. The control switch K_1-K_{2n+1} is on the left side of the battery pack. The corresponding cell is A_1-A_{2n+1}. The control switch K_0-K_{2n} is on the right side of the battery pack and the corresponding cell is A_0-A_{2n}. The control switch network on the left is connected to the output side winding A_{2N-1} of the flyback transformer. The control switch network on the right side is connected to the output side winding A_2, the battery pack is integrally connected to the L_0 end of the flyback transformer.

The flyback mode was applied when the number of unbalanced battery cells in the battery pack increased, which indicated that the energy of the battery pack and the battery cells are directly subjected to two-way energy transfer. There are two kinds of equilibrium situations. One is that the excess energy in the high-power battery cell was transferred to the positive battery pack, eliminating the effect of excessive power. The other is that the energy was provided by the battery cell with low battery capacity, so that the damage of the battery itself induced by over flushing or over discharging of an individual cell can be avoided.

When the energy difference between the individual cells in the battery pack is large, and the equalization mode will adopt forward operation mode, the cell can be directly connected to the A_{2N-1} and A_2 windings on the output side for fast equalization.

The new battery equalizer uses the two modes of operation to work together. Flyback mode is "rough-tuning" and Forward mode is "fine-tuning" synergism, which achieves the goal of efficient and balanced, making efficient use of energy.

2.2. Equilibrium Process Analysis

According to the voltage difference in different single cells as the judgment basis. The equalization method was proposed in this paper to select forward operation mode or flyback operation mode. Finally, efficient and fast equalization of the battery pack was realized, and the specific balancing strategy will be discussed in detail as follows.

Case 1: when the batteries in the group are unbalanced and the voltage difference is large (less than 0.4 V), the operation flyback mode timing is as shown in Figure 6a. In a work cycle, the working process of the flyback transformer can be divided into two stages according to the difference between the control signal and the current path in the circuit. During the process of charging, the energy was transferred from the primary side L_0 of the transformer to the secondary sides L_1 and L_2. The drop of conduction voltage in the control switch is V_{D1} and V_{D2}. The overall primary side voltage V_{L0} of the battery pack and the output side voltage are V_{L1}, V_{L2}. The energy was transferred from the secondary side of the transformer to the primary side during discharge.

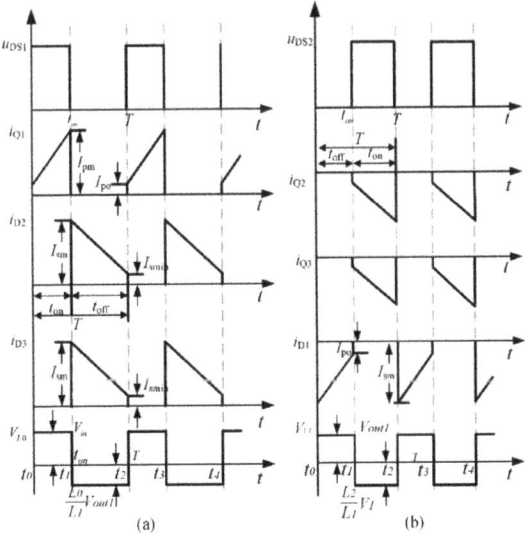

Figure 6. Flyback transition timing diagram. (a) Flyback converter energy forward flow timing diagram; (b) Flyback converter energy reverse flow timing diagram.

The following circuit workings in detail were defined as follows: the leakage inductance of the multi-winding transformer is L_k, the magnetizing inductance denotes Lm. In the primary winding of the transformer, the voltage at the input port is V_{L0}, the output port voltage is V_{L1} and V_{L2}.

$$V_{L1} = -(\frac{N_{L0}}{N_{L0}})V_{L0} \tag{1}$$

$$V_{L2} = -\left(\frac{N_{L2}}{N_{L0}}\right)V_{L0} \tag{2}$$

The coefficient of leakage inductance for V_{L0} is K, which can be expressed as:

$$K = \frac{L_k}{L_m} \tag{3}$$

The energy is transmitted from the primary side to the secondary side, which was shown in Figure 6b. It is divided into the following two modes.

Model 1 ($t_0 \sim t_1$): the Q_1 switch is turned on, the current I_p in the primary side L_0 of the transformer increases. When $t = t_{off}$, I_Q reaches the maximum value. The energy is stored in the magnetizing inductance L_m of the transformer. Where in the exciting inductor current i_{LM} is:

$$i_{Lm} = i_{Lk} = \frac{V_{L0} - V_D}{L_m + L_k} t_{on} \tag{4}$$

The peak value of the inductor current at the time of t_{on} is:

$$i_{pk} = i_{LK}|_{t=t_{on}} = \frac{(V_{L0} - V_D)DT}{L_m(K+1)} \tag{5}$$

where D is the duty cycle of the control switch. T is the turn-on switching period of the flyback converter, the energy stored E in the transformer is:

$$E = \frac{1}{2}L_m i_{pk}^2 = \frac{(V_{L0} - V_D)^2 D^2 T^2}{2L_m(K+1)^2} \tag{6}$$

In this process, the iron core of the transformer is magnetized. Wherein the increase in the magnetic flux Φ is:

$$\Delta\Phi_{(+)} = \frac{V_{L0}}{N_{L0}} D t_{on} \tag{7}$$

Model 2 ($t_1 \sim t_2$): the Q_1 switch is turned off, the secondary side switches Q_2 and Q_3 are both open. Its current is slowly reduced to Ismin, i_{D2} and i_{D3}, charge capacitors C_1 and C_2. The energy in the stored primary side is transferred to the low-energy cell for charge equalization. The secondary inductor current gradually decreases and the falling slope is:

$$\frac{di_{L1k}}{dt} = -\frac{V_{L1} + V_{D1}}{L_1 + L_{1k}} \tag{8}$$

$$\frac{di_{L2k}}{dt} = -\frac{V_{L2} + V_{D2}}{L_2 + L_{2k}} \tag{9}$$

In this process, the core of the transformer is demagnetized. The magnetic flux Φ is also linearly reduced, where the reduction Φ is:

$$\Delta\Phi_{(-)} = \frac{U_{L1}}{N_{L0}}(1-D)t_{on} \tag{10}$$

The process of turning on and off the energy from the secondary side to the primary side, in principle, is the same as the above process. Therefore, the description is not repeated.

Case 2: when the cells in the group are unbalanced, the voltage difference becomes large (no less than 0.4 V). A forward operating mode was applied in the topology without the output of inductance, considering that the selected battery cells A_{2N-1} and A_2 are monomers with a large difference in power. The circuit topology is shown with the dotted line in Figure 5. The specific working process of this model is analyzed as follows. The working sequence diagram is shown in Figure 7.

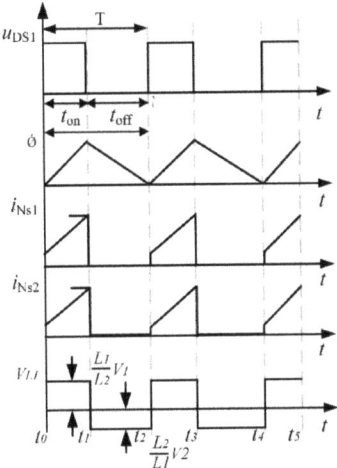

Figure 7. Forward transition timing diagram.

(1) Model 1($t_0 \sim t_1$): when $t = 0$, the switch tubes Q_2 and Q_3 are turned on. The voltages of the battery cells A_{2n-1} and A_2 are applied across the output windings L_1 and L_2 of the flyback converter, respectively. Battery A_{2n-1} charges the primary winding Lm1 through Q_2 as shown in Figure 5. The voltage across Lm is $V_{A_{2n-1}} - V_{D2}$ where the direction is up and down. Meanwhile, the induced voltage across the winding L_{m2} is $V_{A_{2n-1}} - V_{D2}$ due to the coupling relationship of the voltage transformation, its direction is up and down. A peak current of I_0 was produced from the secondary winding circuit under the turn-on instant. The core is magnetized and the magnetic flux is maximized during this process. The current on winding L_1 and winding L_2 increases and the voltage across the battery is directly parallel balanced.

$$I_0 = \frac{V_{L1} - V_{D1} - V_{A2}}{R_{eq}} \quad (11)$$

where R_{eq} denotes the equivalent resistance of the battery A_2 loop, then the voltage across the secondary winding L_2 is clamped at $V_{A2} + V_{D2}$. The inductance current in the circuit decreases linearly.

Its slope is:

$$\frac{di_{L2k}}{dt} = -\frac{V_{A2} + V_{D2}}{L_2 + L_{2k}} \quad (12)$$

During the period of t_0-t_1, the currents of windings L_1 is:

$$i_{L1}(t) = I_0 - \frac{V_{A2N-1} + V_{D1}}{L_1 + L_{1K}}(t - t_0) \quad (13)$$

At time t_1, the bidirectional switches Q_1 and Q_2 are turned off. The mode ends and the current of the winding L_1 at time t_1 is:

$$i_{L1}(t1) = I_0 - \frac{V_{A2N-1} + V_{D1}}{L_1 + L_{1K}}DT \quad (14)$$

The current of the primary winding at time t_1 is:

$$i_{L1k}(t_1) = I_{max} + \frac{V_{A2N-1} - V_{D1}}{L_{1m}(1+K)}DT \quad (15)$$

The energy stored in the transformer at this time is:

$$W_{L1m} = \frac{(V_{A2N-1} - V_D)^2}{2L_{1M}(1+K)^2} D^2 T^2 \tag{16}$$

(2) Model 2 ($t_1 \sim t_2$): at $t = t_{on}$ time. Turn off the switch tubes Q_2 and Q_3. Switch tubes Q_2 and Q_3 is turned off when t is at t_{on}. There is no current flowing through the winding L_1 and L_2. At this point, the transformer conducts magnetic reset through the reset winding. The excitation current is fed back to the whole battery pack through the primary side. Finally, the reset of the magnetic flux is realized.

$$\frac{di_{L2k}}{dt} = -\frac{V_{A2} + V_{D2}}{L_2 + L_{2k}} \tag{17}$$

$$i_{L2}(t) = I_{L2}(t_1) - \frac{V_{A2} + V_{D2}}{L_2 + L_2 K}(t - t_1) \tag{18}$$

At this point, the energy is completely transferred at time t_2. The current $i_{L2}(t)$ becomes zero and the transformer completes the magnetic reset.

2.3. Design Consideration of the Main Circuit

This article uses a two-winding bidirectional flyback converter. Whether energy is from V_{in} to V_{out}, or energy is from V_{out} to V_{in}, the switch Q_1 is turned on when the diode D_1 opposite to it is energized. The switch tube Q_2 is turned on. When the D_2 is energized and all the zero-voltage switching (ZVS) are turned on. In the bidirectional flyback converter, the current alternates and the current work continuously. Let Q_1 have a duty cycle of D_{Q1}. When the energy flows V_{in} to V_{out1}, the input voltage and output port voltage are expressed as:

$$\frac{V_{IN}}{V_{out1}} = \frac{L_2}{L_1} \times \frac{D_{Q1}}{1 - D_{Q2}} \tag{19}$$

If D_{Q2} is the duty cycle of Q_2, when energy flows from V_{out1} to V_{in1}:

$$\frac{V_{out1}}{V_{in1}} = \frac{L1}{L2} \times \frac{D_{Q2}}{1 - D_{Q1}} \tag{20}$$

Assuming that the input voltage Vin is 30–45 V, the output voltage Vout is 3.8–5 V and the switching period T is 40 μs. And then the duty ratio D can be derived to be 68%. The average equalization current I_{pk} is calculated to be 1.52 A.

The peak equalization current is expressed as follows:

$$I_{pk} = \frac{2P_{in}}{D_{max} V_{inmin}} \tag{21}$$

The input power is expressed as follows:

$$P_{in} = \frac{p_0}{E_{ff}} = \frac{U_{out1} \times I_{w1} + U_{out2} \times I_{w2}}{E_{ff}} \tag{22}$$

Considering the energy storage and conversion of the flyback converter. The primary inductance is expressed as follows:

$$L_m = \frac{V_{inmin} \times D_{max}}{I_{pk} \times f_s K_{RF}} = \frac{V_{inmin} \times D_{max}}{\left(\frac{2P_{in}}{D_{max} V_{inmin}}\right) \times f_s K_{RF}} = \frac{(V_{inmin} \times D_{max})^2}{2P_{in} f_s K_{RF}} \tag{23}$$

The flyback transformer is equivalent to a coupled inductor. In the process of transferring energy from the primary side to the secondary side, because the coil has magnetic flux leakage the magnetic flux leakage energy cannot be transferred to the secondary party thereby causing the switching tube voltage to rise further. The clamp circuit connected to the primary and secondary sides of the bidirectional flyback transformer. This limits the switching tube voltage to the maximum value. This article uses the RCD (Resistance Capacitance Diode) absorption circuit, as shown in Figure 8.

$$R_C = \frac{V_C^2}{P_R} = \frac{2 \times V_C^2}{L_{ik} I_p^2 f_s} \quad (24)$$

$$C_C \geq \frac{V_C}{\Delta V_C R_C f_s} \quad (25)$$

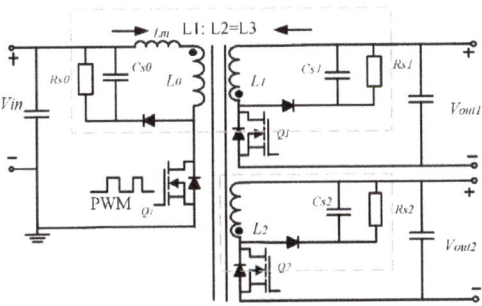

Figure 8. RCD absorption circuit.

The clamp capacitor C has the same voltage as the resistor R. The relationship between R and C can be obtained by the relationship between voltage and power transition. According to the above analysis, during the working process of the clamp circuit, the energy stored in the capacitor is consumed by the resistor. That is, the power consumed by the resistor is equal to the energy stored by the capacitor.

2.4. Equalization Strategy

The new voltage equalization structure was proposed in this paper. During the process of equalizing the battery pack, it is necessary to control the DC–DC converter in different operating modes. This can become an essential basis for working mode matching by the amount of cell voltage difference. Meanwhile, the corresponding control method was adopted to control the equalization switch network. The balancing strategy is shown in Figure 9. The strategy of equalization control should pay attention to the following points:

A. Selected battery port voltage as a criterion for equalization.

B. In the process of energy transfer, when the current is not zero, it will cause irreversible losses to the elements of the switch array in the long-term equalization process.

C. Balanced operation modes are divided into two types, which are switched according to the distribution of different battery voltages.

For A: the equalization control uses the battery port voltage as a reference. The equalization mode is selected according to the voltage dispersion degree of different unit cells. The highest voltage and the lowest voltage of the battery cells are V_{max} and V_{min}, and the difference between them is V_d. The difference between the voltages is:

$$V_d = V_{max} - V_{min} \quad (26)$$

For B: when the switch network is switching. It is necessary to keep the flyback isolation converter off. Prevents damage to circuit components, which is caused by inrush current during switching.

For C: when the allowable difference V_d set by V_d is equalized, the balance is stopped until V_d is less than V_d. The specific equalization process is played by two working modes, which will be described in detail below as follows.

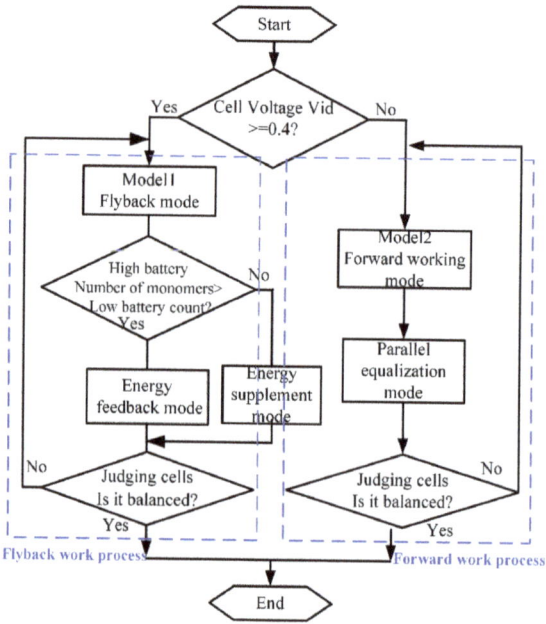

Figure 9. Equalization strategy flow chart.

Flyback mode: the problem of the voltage difference V_d has been addressed when this difference of each battery in the battery pack is no less than 0.4V. The battery pack as a whole was applied as energy input and only the battery pack was used as energy output through the flyback mode of operation. Due to the special two-winding output structure. The battery pack can simultaneously charge balance two low-power batteries, or two high-energy batteries can be fed as energy input for the entire battery pack.

Forward mode: when the pressure difference V_d of each cell in the battery pack is less than 0.4V, in forward mode select two low-power battery cells by connecting the two output windings of the transformer. Point-to-point supplementation of energy for fast balancing purposes.

3. Experimental Results

3.1. Voltage Equalization System

Figure 10 shows the structure of the voltage equalization system designed by the new voltage equalization structure proposed in this paper. Because the car power battery pack is composed of series batteries. Therefore, this article uses 12 battery cells in series for voltage equalization experiments. the battery voltage of the series can be monitored in real-time using the battery detection chip LTC6803. The main control unit communicates with the LTC6803 through SPI (Serial Peripheral Interface) to obtain the voltage of the battery cells of the battery pack and control the LTC6803. The master passes the output control signal. The LTC6803 controls the two-sided interleaved switching network through the S1, S2, S3...S12 pins. The master provides control and drives signals for the flyback isolation converter.

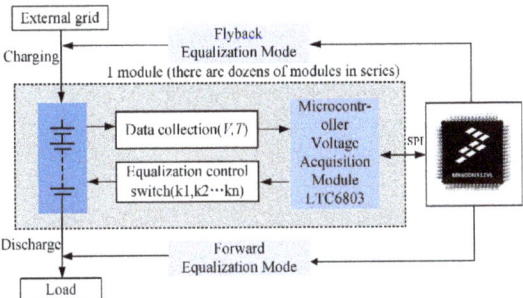

Figure 10. Voltage equalization system.

3.2. Main Circuit Design

In the new voltage equalization system, real-time acquisition and analysis for parameters of battery voltage and temperature parameters were performed by the main controller and voltage detection chip LTC6803. Different equalization modes are selected according to the number of unbalanced monomers, thus the equalization control was efficiently implemented. The switch array has a function of selecting a path, and the battery cells are connected to the double-winding flyback isolating converter by the switches on the left and right sides. The switch consists of a bidirectional MOS (Metal-Oxide-Semiconductor) tube. The six cells A1, A2, A3, A4, A5 and A6 at the bottom of the pool group are connected to the N-MOS (N-Metal-Oxide-Semiconductor) switch tube respectively. The left battery cells A7–A12 are connected to the P-MOS (P-Metal-Oxide-Semiconductor) switch tube, the specific working circuit is shown in Figure 11. The battery cell is connected to the switch tube through the optocoupler, the switch is connected to the gate of four MOS transistors, the first four switches of the battery pack are Sa12, Sb12, Sa11, Sb11, the gate of each set of switches is connected to the c-pole of the phototransistor in the optocoupler, the e pole is connected to the negative electrode of the battery cell, which possessed the lower three positions of the battery cell. For example, the switch corresponding to the battery unit A12 is Sa12. The gate of the four PMOS transistors is connected to the c pole of the optocoupler OPa12, and the e pole is connected to the negative electrode of the battery unit A9. The switch corresponding to the A1 battery cell is OPb1 on the right side in the switch array, the gates of the four N-MOS transistors are connected to the e-pole of the optocoupler OPb. The c pole is connected to the battery cell A4. The connection method is the same as the above discussion when the control switches of the remaining battery units are applied. An additional complicated switch drive circuit is not required for the structure, the driving voltage of MOS is directly provided by the battery pack, the optocoupler is connected to the left and right switch arrays for signal control, which acts as a signal isolation function, the battery structure is simple and efficient.

The circuit of the flyback isolation converter is shown in Figure 12. The output winding of the flyback converter is connected to two battery cells of different potentials in the battery pack. The corresponding equalization mode was adopted through different modes. The isolation between the battery pack and battery can be realized by the flyback transformer T. The optocoupler OPT, MOS transistor S3 and S4 formed an isolation driver. The 5V power supply was provided by an isolated power supply. Current is detected by current sensor ACS712. The control switches in the circuit are Q_1, Q_2, Q_3 respectively. If switch Q_1 is turned on, meanwhile Q_2 and Q_3 are off, and the duty cycle PWM pulse can be adjusted by the main controller according to the output of the equalization circuit. The signal that controls the output was divided three ways: one way was to control V_{g1} and the two external signals were connected to one inverter to control the opposite signals of V_{g2} and V_{g3}. The driving circuits of Q_2 and Q_3 shown in Figure 12 are the same as the working principle, which thus was simplified in the circuit diagram.

When the input signal V_g is high, the LED (Light Emitting Diode) in the optocoupler did not work. The phototransistor is turned off, resulting in a high-level output of optocoupler. Drive control MOS tube S3 is turned on, and S4 is turned off. The output voltage of the isolated power supply is applied for the gate level of the switch Q_1 through the resistor R_3. When it turns off, a small amount of charge left to generate a voltage difference with the S4 gate voltage, S4 is turned on due to the influence of the Q_1 gate capacitance. Then the Q_1 gate is discharged, and Q_1 is turned off. The main component parameters of the system are shown in Table 1.

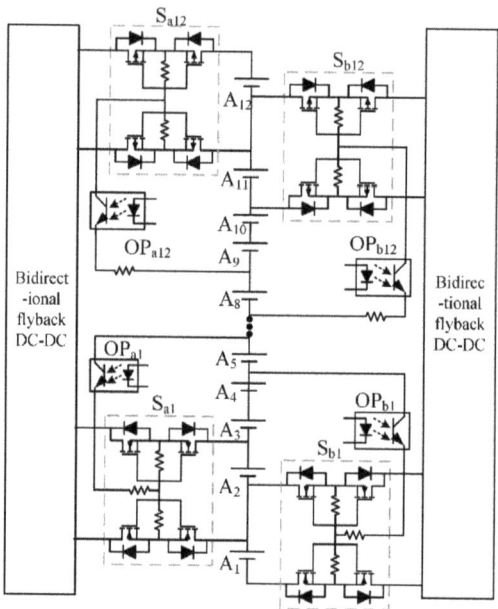

Figure 11. Switching network diagram.

Table 1. Main component parameters of equalizing circuit system.

		Parameter		Value
Primary DC–DC Converter		Mosfect Switch		IPB200N
		Rectifier Diode		IN5822
	Transfor-mer	Core		EP20
		N1:N2:N3		33:3:3
		Lm		367 uH
	Primary RCD	R		38 kΩ
		C		21 nF
Secondary DC–DC converter		Mosfect Switch		AUIRF3504
		Rectifier diode		SR560
	Secondary RCD	R		803 Ω
		C		1 uF
Selection Switch		P-MOS		AOD409
		N-MOS		AUIRF3504
		Optocupler		PC817

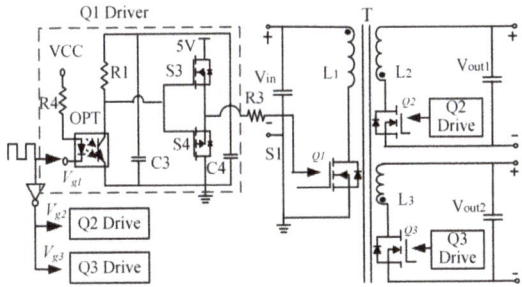

Figure 12. Flyback switch drive circuit.

3.3. Experimental Verification

To verify the performance of the proposed voltage equalizer, a prototype was built and tested. Figure 13 shows a photograph of the experimental prototype and associated instruments. The experiment uses a "DELI PU BATTERY" 18650 Li-ion Battery with a capacity of 2800-mAh, a rated voltage of 3.7 V, a discharge cut-off voltage of 2.75 V, and an equivalent series resistance of 80 mΩ. The standard charging voltage is 0.5 C (1400 mA), the operating temperature is 0–45 °C in the state of charge, and the discharge state is −(20–60) °C. the battery pack consists of 12 batteries of 2800-mAh battery strings which can store 10.36 Ah and 124.32 Wh. When the experiment is carried out, the voltage detection frequency is 50 Hz. The switching frequency of the left and right switches is 1 Hz, the operating frequency of the bidirectional winding flyback converter is 25 kHz and the voltage difference between the beginning and the ending equalization is 30 mV.

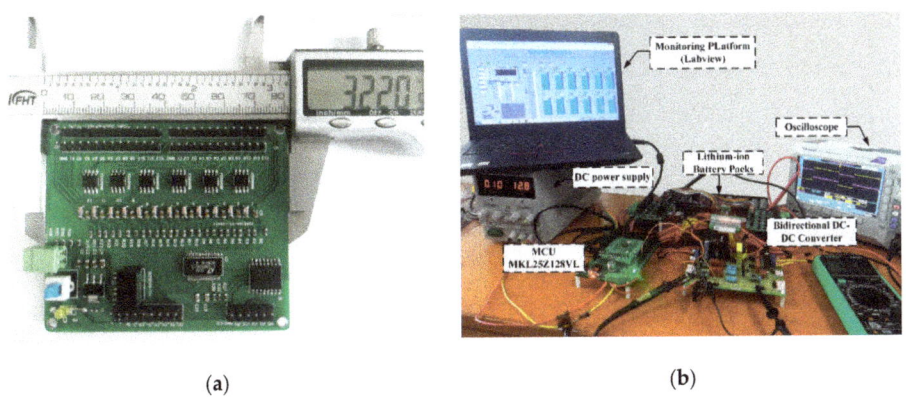

(a) (b)

Figure 13. (a) Voltage acquisition circuit; (b) experimental prototype and associated instruments.

Figure 14 shows the voltage variation of the battery cells during the overall equalization process of the battery pack. It can seem that the highest and lowest voltage of the single battery was 3.98 V and 2.78 V respectively. The entire energy transfer and conversion process can be realized by bidirectional transfer, and a gradual reduction in high voltage cells was achieved. The low voltage gradually rises and cycles in sequence until all voltages were up to the set equilibrium voltage.

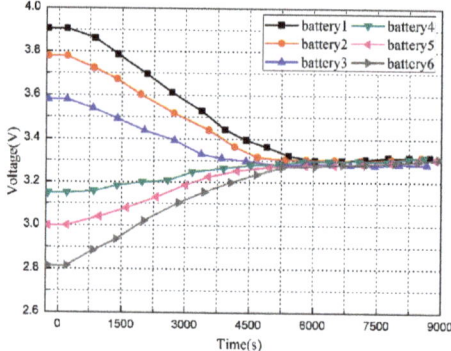

Figure 14. Voltage curve during forward go equalization.

Figure 15 shows a schematic diagram of voltage changes during battery pack equalization. In this mode of operation, 12 cells with large energy differences are selected for direct equalization. From the analysis of the graph, a relatively balanced state can be gradually obtained between the highest-energy battery cells and the lowest-connected battery cells, and the remaining energy was also in accordance with this equalization mode.

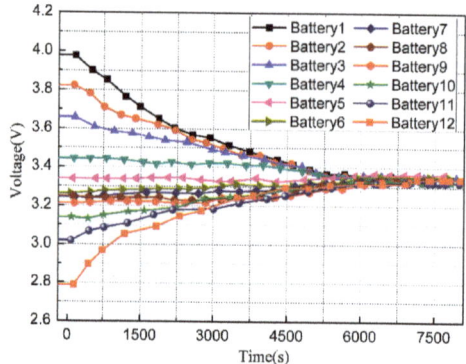

Figure 15. Voltage curve during equalization process.

Figure 16 shows the voltage distribution before and after the battery voltage equalization. The horizontal axis represents different monomers and the vertical axis denotes the cell voltage. It can be noted clearly that the voltage of each cell in the battery pack is unbalanced. The voltages tend to be consistent upon equalization, which indicates that the new voltage equalization structure can effectively realize the function of voltage equalization.

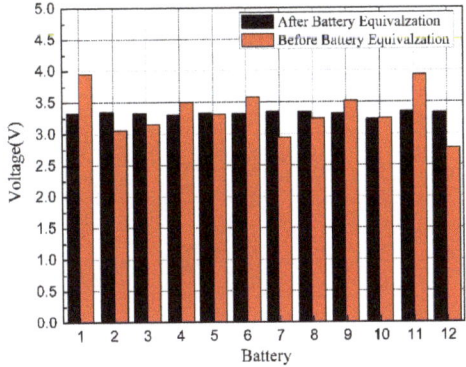

Figure 16. Voltage distributions of the battery pack.

4. Discussion

A comparative study of the proposed equalization method is shown in Table 2. Through comparing and analyzing the new equalizer with other equalization networks, the proposed equalizer has relative advantages in size, cost and equalization rate. The output structure of double windings is utilized to realize efficient operation in two working modes of forward and reverse excitation. It is suitable for a large-scale battery

A pack-to-cell equalization mode was formed, which has great potential in promotion and application [14]. With the coaxial winding structure, the equalization function can be realized without additional control in the whole process, but the disadvantage is that it is difficult to accurately adjust the voltage of the battery pack in actual work. The parasitic inductance and mutual inductance of the circuit cannot be accurately controlled, which influenced the operation of the circuit. This leads to the difficult consistency of battery packs for the battery management system (BMS). Additionally, this structure is not suitable for a large number of battery cells. It is challenging to wind too many windings on the same core. Finally, the true energy balance of the battery pack made it difficult to achieve balance from the battery pack to the single unit [15]. It cannot directly reduce the monomer with the highest energy and directly supplement the monomer with the lowest energy. Energy can only be transferred from the whole battery pack.

Energy can only be transferred in the whole battery pack one by one during the equalization process. However, the equalization speed is relatively slow for most batteries connected in series. A point-to-point balanced supplement can be realized by the proposed point-to-point balanced structure, but the number of switch groups is huge. If a small pressure difference has existed between the two equalization monomers in the equalization process, the equalization time is not dominant. The equalizer is based on the equalization of coaxial windings. At the same time, the advantages of a high-frequency switch of the forward converter and the small volume of flyback converter were retained. The number of switches was relatively small and the control method was flexible. In addition, in order to prevent the surge impact of high voltage and high current on the cells during the charging and discharging process of the pack, a Zener diode with an anti-parallel protection function on every single cell was applied in the new equalization structure, and the battery pack module was also equipped with fuses to ensure the safe operation of the battery pack. the proposed equalizer exhibited higher stability and better equalization by this method.

Table 2. Parameters the prototype.

Balanced Mode	Energy Dissipation Type		Non-Energy Dissipation Type							
			Cell to Cell		Pack to Vell		Cell to Pack	Cell to Pack to Cell		
Type	Fixed Resistor [7]	Shut Resister [8]	Single Layer [9]	Buck-Boost [12]	Multi-Stage Winding [14] Trans-Former	Single Winding [15] Trans-Former	Many-to-One [16]	Two Way [17]	Inter Group [18]	#New Proposed Topology
Inductance	0	0	1	N	0	0	N	0	2n+4	0
Capacitance	0	0	1	2N	1	1	0	0	0	3
Transformer	0	0	0	0	2	1	N	2	2	1
Switch	0	N	2N+4	N	4	2N	N	2N	N	2N
Diode	0	0	0	0	N+1	1	N	1	0	0
Volume	E	E	E	E	P	G	P	P	P	G
Control method	E	E	P	S	P	P	E	G	G	E
speed	P	G	G	G	G	G	E	G	G	E
cost	E	E	E	E	P	P	P	P	P	G

E: excellent, G: good, S: satisfactory, P: poor.

5. Conclusions

In order to solve the problem of inconsistency between the single cell in the actual use process of the power lithium-ion battery pack, this paper displays a new type of voltage equalization circuit. Comparing with the traditional equilibrium structure, the balanced mode adopted was divided into two working modes. The control method is flexible and the volume occupied by the space volume is relatively small. The balanced structure of the two windings could allow the fast equalization of two battery cells simultaneously. Meanwhile, the designed switch array circuit has a simple control mode with low energy loss. In this paper, the working principle and equalization control strategy of the voltage equalization circuit was introduced in detail and analyzed. A voltage equalization system for 12 batteries of 2800-mAh battery strings was designed in a targeted manner and a voltage equalization experiment was performed. The results show that voltage equalization of series battery packs can be realized efficiently and quickly by the proposed equalizer. The circuit became simple and efficient using a flexible control mode. It is beneficial to improve the overall performance of the battery pack, meanwhile guaranteeing the healthy operation of the battery pack.

Author Contributions: F.S. designed the study, D.S. designed the experiments and wrote the manuscript. All authors have read and approved the manuscript.

Funding: Research funded by National Natural Science Foundation of China (57807138).

Acknowledgments: Thank you for the experimental instrument support and financial support provided by the Engineering Training Center of Tianjin University of Technology.

Conflicts of Interest: This article has no conflict of interest.

References

1. Preindl, M.A. Battery Balancing Auxiliary Power Module with Predictive Control for Electrified Transportation. *IEEE Trans. Ind. Electron.* **2018**, *65*, 6552–6559. [CrossRef]
2. Kouchachvili, L.; Yaïci, W.; Entchev, E. Hybrid battery/supercapacitor energy storage system for the electric vehicles. *J. Power Sources* **2018**, *374*, 237–248. [CrossRef]
3. Cao, X.; Zhong, Q.C.; Qiao, Y.C.; Deng, Z.Q. Multi-layer Modular Balancing Strategy for Individual Cells in a Battery Pack. *IEEE Trans. Energy Convers.* **2018**, *33*, 526–536. [CrossRef]
4. Gallardo-Lozano, J.; Romero-Cadaval, E.; Milanes-Montero, M.I.; Guerrero-Martinez, M.A. Battery equalization active methods. *J. Power Sources* **2014**, *246*, 934–949. [CrossRef]
5. Lee, K.M.; Lee, S.W.; Choi, Y.G.; Kang, B. Active Balancing of Li-Ion Battery Cells Using Transformer as Energy Carrier. *IEEE Trans. Ind. Electron.* **2017**, *64*, 1251–1257. [CrossRef]
6. Lee, K.M.; Chung, Y.C.; Sung, C.H.; Kang, B. Active Cell Balancing of Li-Ion Batteries using LC Series Resonant Circuit. *IEEE Trans. Ind. Electron.* **2015**, *62*, 5491–5501. [CrossRef]
7. Kutkut, N.H.; Divan, D.M. Dynamic equalization techniques for series battery stacks. In Proceedings of the Intelec'96—International Telecommunications Energy Conference, Boston, MA, USA, 6–10 October 1996; pp. 514–521.
8. Lindemark, B. Individual cell voltage equalizers (ICE) for reliable battery performance. In Proceedings of the Thirteenth International Telecommunications Energy Conference—INTELEC 91, Kyoto, Japan, 5–8 November 1991; pp. 196–201.
9. Park, S.-H.; Kim, T.-S.; Park, J.-S.; Moon, G.-W.; Youn, M.-J. A new battery equalizer based on buck-boost topology. In Proceedings of the 7th International Conference on Power Electronics, Daegu, Korea, 22–26 October 2007; pp. 962–965.
10. Elserougi, A.; Massoud, A.; Ahmed, S. A Unipolar/Bipolar High-Voltage Pulse Generator Based on Positive and Negative Buck-Boost DC-DC Converters Operating in Discontinuous Conduction Mode. *IEEE Trans. Ind. Electron.* **2017**, *64*, 5368–5379. [CrossRef]
11. Li, S.; Mi, C.C.; Zhang, M. A High-Efficiency Active Battery-Balancing Circuit Using Multiwinding Transformer. *IEEE Trans. Ind. Appl.* **2013**, *49*, 198–207. [CrossRef]

12. Shang, Y.; Zhang, C.; Cui, N.; Guerrero, J.M. A Cell-to-Cell Battery Equalizer with Zero-Current Switching and Zero-Voltage Gap Based on Quasi-Resonant LC Converter and Boost Converter. *IEEE Trans. Power Electron.* **2015**, *30*, 3731–3747. [CrossRef]
13. Li, Y.; Han, Y. A Module-Integrated Distributed Battery Energy Storage and Management System. *IEEE Trans. Power Electron.* **2016**, *31*, 8260–8270. [CrossRef]
14. Xiong, H.; Fu, Y.; Dong, K. A novel point to point energy transmission voltage equalizer for Series-Connected Supercapacitors. *IEEE Trans. Veh. Technol.* **2016**, *65*, 4669–4675. [CrossRef]
15. Karnjanapiboon, C.; Jirasereeamornkul, K.; Monyakul, V. High efficiency battery management system for serially connected battery string. In Proceedings of the IEEE International Symposium on Industrial Electronics, Seoul, Korea, 5–8 July 2009; pp. 1504–1509.
16. Kutkut, N.H.; Wiegman, H.L.N.; Divan, D.M.; Novotny, D.W. Design considerations for charge equalization of an electric vehicle battery system. *IEEE Trans. Ind. Appl.* **1999**, *35*, 28–35. [CrossRef]
17. Imtiaz, A.M.; Khan, F.H.; Kamath, H. A Low-Cost Time Shared Cell Balancing Technique for Future Lithium-Ion Battery Storage System Featuring Regenerative Energy Distribution. In Proceedings of the Twenty-Sixth Annual IEEE Applied Power Electronics Conference and Exposition (APEC), Fort Worth, TX, USA, 6–11 March 2011; pp. 792–799.
18. Lim, C.S.; Lee, K.J.; Ku, N.J.; Hyun, D.S.; Kim, R.Y. A Modularized Equalization Method Based on Magnetizing Energy for a Series-Connected Lithium-Ion Battery String. *IEEE Trans. Power Electron.* **2014**, *29*, 1791–1799. [CrossRef]

 © 2019 by the authors. Licensee MDPI, Basel, Switzerland. This article is an open access article distributed under the terms and conditions of the Creative Commons Attribution (CC BY) license (http://creativecommons.org/licenses/by/4.0/).

Article

Semiactive Hybrid Energy Management System: A Solution for Electric Wheelchairs

Sadam Hussain, Muhammad Umair Ali, Sarvar Hussain Nengroo, Imran Khan, Muhammad Ishfaq and Hee-Je Kim *

School of Electrical Engineering, Pusan National University, Busandaehak-ro 63beon-gil, Geumjeong-gu, Busan 46241, Korea; sadamengr15@gmail.com (S.H.); umairali.m99@gmail.com (M.U.A.); ssarvarhussain@gmail.com (S.H.N.); imrankhanyousafzai4159@gmail.com (I.K.); engrishfaq1994@gmail.com (M.I.)
* Correspondence: heeje@pusan.ac.kr; Tel.: +82-51-510-2364

Received: 21 February 2019; Accepted: 20 March 2019; Published: 21 March 2019

Abstract: Many disabled people use electric wheelchairs (EWs) in their daily lives. EWs take a considerable amount of time to charge and are less efficient in high-power-demand situations. This paper addresses these two problems using a semiactive hybrid energy storage system (SA-HESS) with a smart energy management system (SEMS). The SA-HESS contained a lithium-ion battery (LIB) and supercapacitor (SC) connected to a DC bus via a bidirectional DC–DC converter. The first task of the proposed SEMS was to charge the SA-HESS rapidly using a fuzzy-logic-controlled charging system. The second task was to reduce the stress of the LIB. The proposed SEMS divided the discharging operation into starting-, normal-, medium-, and high-power currents. The LIB was used in normal conditions, while the SC was mostly utilized during medium-power conditions, such as starting and uphill climbing of the EW. The conjunction of LIB and SC was employed to meet the high-power demand for smooth and reliable operation. A prototype was designed to validate the proposed methodology, and a comparison of the passive hybrid energy management system (P-HESS) and SA-HESS was performed under different driving tracks and loading conditions. The experimental results showed that the proposed system required less charging time and effectively utilized the power of the SC compared with P-HESS.

Keywords: electric wheelchair; lithium-ion battery; supercapacitor; semiactive hybrid energy storage system; smart energy management system

1. Introduction

Currently, there are an estimated 600 million people aged 60 years or older in the world [1]. In addition, people disabled due to traffic and lower-limb accidents add another 9 million to the count, with an increasing rate of 500,000 per year. The quality of life of elderly or disabled people is restricted. However, advancements in different assistive devices, such as wheelchairs, has led to an increase in their range of activities [2]. While using electric wheelchairs (EWs), people want to travel greater distances and reduce the amount of time it takes to charge the battery [3].

Various technologies have been employed for EWs, but their efficiency greatly depends on the characteristics of their energy storage system (ESS) [4,5]. Various ESSs, including lithium-ion batteries (LIBs), lead–acid batteries, and nickel metal hydride batteries, are used in vehicular applications [6,7]. Among these, LIBs are a widely used energy source due to their attractive properties such as high energy density, low self-discharge rate, and long lifecycle [8–10]. A comparison of different properties of batteries is shown in Figure 1. On the other hand, in vehicular applications, the battery faces many challenges, such as the need for high power demand during acceleration or uphill climbing modes. Although a high-power battery is possible to tackle this problem, these batteries are quite bulky and expensive [11,12]. The power specifications of the LIB are very low, and the peak-to-average-power ratio ranges between 0.5 to 2 [13], which makes LIBs unpromising in high-power-demand situations. Therefore, supercapacitors (SCs) can be used as a secondary ESS.

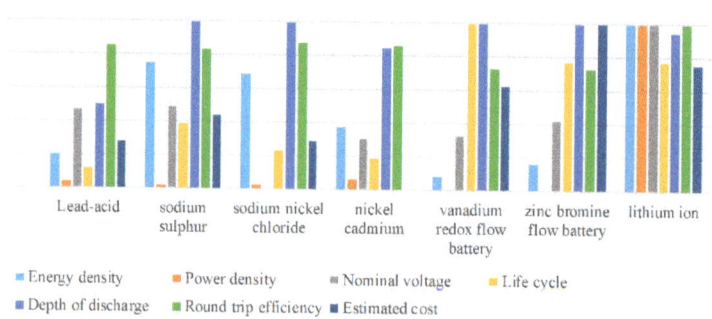

Figure 1. Comparison of different battery storage systems [14,15].

The power and energy density of SCs range from 1000 to 5000 W/kg and 1 to 10 Wh/kg, respectively [16]. SCs have charging and discharging capabilities of 10–20 times more than LIBs [17]. SCs are not a replacement for LIBs but provide the power needed for EW systems when accelerating up a slope. In addition, they have a longer lifecycle and a wider range of operating temperatures (−4 to +70 °C) than LIBs, although they do have low energy density [18]. In electric vehicles (EVs), the stress on the battery is higher than that in a hybrid electric vehicle (HEV) due to deep discharging (~80% for EVs and ~10% for HEVs) [19]. A hybrid energy storage system (HESS) is a combination of SC and LIB, which combines the advantages of both devices to fulfill the requirements of high energy and power densities. SCs are normally used for high power storage, and LIBs are used as a high-energy-storage unit. The SC is utilized for the high power demand of the powertrain, while the LIB is used in low-power situations [20].

An energy management system (EMS) is required to fully utilize the energy of both storage units effectively [21]. The EMS presented in [22,23] is based on the frequency decoupling method to protect the battery from abrupt changes in the load. HESSs can be easily divided into three main topologies: passive HESS (P-HESS), semiactive HESS (SA-HESS), and fully active HESS [24]. A P-HESS is the direct coupling of two or more energy storage devices without a power converter [25]. This has several benefits over a standalone LIB as a power source (e.g., higher peak power capability, higher efficiency, and long battery lifecycle) [26]. Although it is simple to implement, there are some limitations. Power sharing is uncontrollable because the two storage systems are not decoupled [27,28]. Napoli et al. [17] used an ultracapacitor connected in parallel to a battery with no power converter between the two sources. In a P-HESS, power sharing between the LIB and SC is determined by their respective resistance, and the resulting terminal voltage follows the discharge curve of a battery [16]. The energy storage device should be decoupled for efficient operations with respect to its characteristics [29]. In contrast, the degree of controllability is increased using a fully active HESS, but there are some disadvantages, such as increased system losses, weight, and cost. In addition, efficiency also decreases because of the additional converter [20]. Similarly, a modular multilevel fully active HESS also adds complexity and cost. This will also affect the sensitivity of the system, with a failure of one DC–DC converter causing system failure [30]. The SA-HESS is a tradeoff between cost and performance. This system employs only one DC–DC converter and most control strategies can be implemented on this topology [31].

The cell voltage of an LIB is very low. Therefore, a string of batteries is generally used. An imbalanced cell voltage because of the series connection of LIBs causes an increase in temperature, deep discharging, overcharging, and reduced lifecycle and capacity of the battery [32]. Different engineering techniques have been used to control and monitor battery parameters (e.g., state of charge (SOC), voltage, current, and temperature) [33]. For voltage balancing, a special type of charging circuit is required [34,35]. Various fast-charging techniques are used to charge LIBs, such as constant current and constant voltage (CC/CV), a multistage current charging algorithm, model-based charging, a pulse charging algorithm, and a fuzzy logic controller (FLC) [36–38]. The CC/CV technique is simple and computationally efficient [39]. This method has two modes. First, it reaches the defined voltage level while providing a CC to the battery. In the second step, a CV is supplied, and the battery current begins to decrease exponentially [40]. This procedure requires a high current if the battery needs to be charged in a short time, but this increases the battery temperature dramatically, which reduces the lifecycle. In contrast, it takes more time to charge a battery if the current is lower. Huang et al. introduced different intervals to determine the optimal current using FLC [41].

This paper proposes a semiactive hybrid energy management system comprised of SA-HESS and a smart energy management system (SEMS). The proposed methodology charges the HESS smartly using an FLC. The temperature is used as a feedback, which is thermally and electrically favorable to achieve a long lifetime for the ESS. The proposed SEMS discharges the SA-HESS smartly according to the desired load current. Several experiments have been performed to compare the proposed technique with P-HESS.

The rest of the paper is organized into five sections. Section 2 describes an overview and the methodology of the proposed system. Section 3 presents the experimental installation. Section 4 reports the results from the experimental study. Section 5 is the discussion of the results, and Section 6 concludes the paper.

2. Methodology

2.1. Overview of the Proposed Methodology

This section provides an overview of the proposed SA-HESS for EWs. At high power/current demand, the use of only an LIB is ineffective and the LIB discharges very rapidly because of the lower power density. Therefore, in the proposed technique, an SC was used as a parallel controlled power source in the SA-HESS. The required power was determined according to the SOC/voltage of the SC and LIB. Figure 2 presents the proposed system for EWs. In the SA-HESS, the LIB was used as a high energy unit, which was connected in parallel to the load via a bus link. The SC was used as a high-power unit, which was connected to the DC bus via a bidirectional converter. The CC/CV charging system was replaced with a temperature (T_{bat}) feedback FLC charging system. The controlled current ($I_{controlled}$) was supplied as the optimal charging current to the LIB. The SC and LIB provided the desired current to the load, which was controlled using an Arduino microcontroller. A bidirectional converter was used in boost mode, whereas discharging and buck mode were used in charging mode [42]. The output voltage of SC (V_{SC}) varied according to its state of charge, while the battery voltage (V_{bat}) remained almost constant. The boost converter was used to maintain the V_{SC} relative to the reference (bus/battery) voltage. The converter generated the pulse width modulation (PWM) signal according to the reference value [43]. The load current (I_L) was sensed using a current sensor and applied as a feedback signal for the SC (I_{SC}) and LIB (I_{bat}) current. The I_{SC} and I_{bat} smartly contributed according to the requirement of I_L.

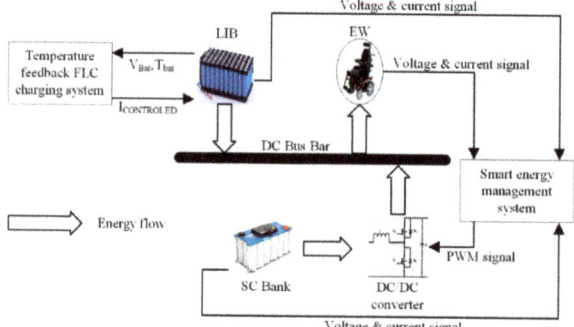

Figure 2. Overview of the proposed methodology. Battery voltage (V_{Bat}), controlled current ($I_{controlled}$), battery temperature (T_{bat}), lithium-ion battery (LIB), electric wheelchair (EW), pulse width modulation (PWM) and supercapacitor (SC).

2.2. Fast-Charging System for EW

FLC is well suited to anticipating a battery's nonlinear behavior because it is robust, easily adaptive, and does not require any mathematical model. The FLC is classified into four parts [44]: Fuzzifier—in the fuzzifier, linguistic fuzzy sets are obtained from the truth value of the membership function. Fuzzy rule base—the fuzzy rule base is designed from professional experience and controls the system operation. Fuzzy interface engine—the fuzzy linguistic input is transformed into a fuzzy linguistic output with respect to the controlled law stated in the fuzzy rule set. Defuzzifier—this maps the fuzzy output from the inference engine to a crisp or real value by using membership functions. The fast-charging methodology was designed using the same rule base as discussed in a previous study [45]. However, in this work, the FLC-based, fast-charging methodology was designed for a series and parallel cell combination of an LIB pack. The lowest single-cell voltage and the highest voltage difference between the two cells of the string were the inputs of the FLC to find the optimal value of the charging current. When the voltage difference between the two cells was high, then a high current was inserted to charge the LIB pack in less time.

However, at the same time, it was essential to control the temperature in order to ensure that this high current value did not affect the battery life. The threshold value and operation of the temperature control unit are shown in Figure 3. The temperature threshold value was set to 39 °C. If the temperature was below the threshold value, $I_{controlled}$ was supplied to the battery. When it crossed the threshold value, then the controller compared I_{bat} with the range defined in the flowchart. The value of the charging current changed according to the value shown in the flowchart.

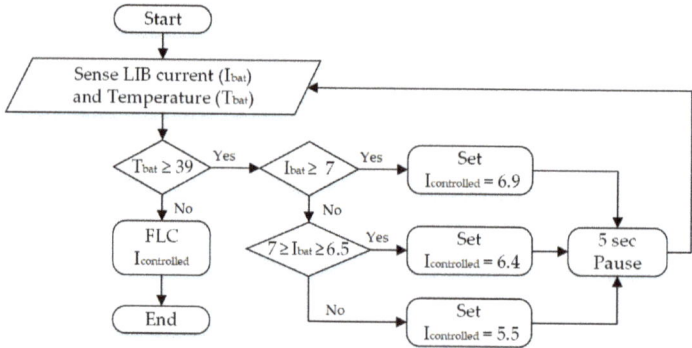

Figure 3. Flowchart for temperature control.

2.3. Smart Energy Management System for the SA-HESS

The power requirement for EWs motion is totally different when traveling on a flat surface than when moving on an inclined surface. A SEMS was designed to overcome the power demand. Two tracks were taken into consideration to implement the SEMS on the SA-HESS. Figure 4a,b present the normal plain track-AB and inclination track-ABCD, respectively. On track-AB, the motor of the EW did not require much power and current; however, for track-ABCD, the motor required high current/power.

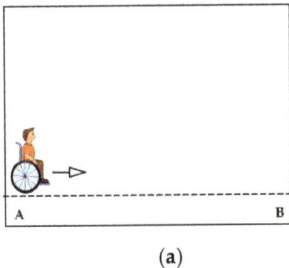
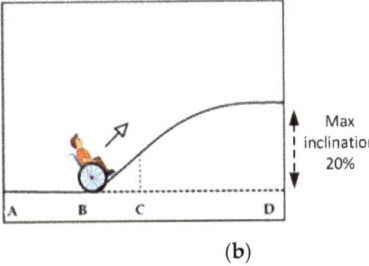

Figure 4. The logic for the smart energy management system (SEMS): (**a**) normal plaint track-AB; (**b**) uphill climbing steep track-ABCD.

The load on the EW's motor can be calculated by using the following equations [46,47]:

$$F = ma - f_x - mg \sin \Theta \qquad (1)$$

$$P = \frac{F \times v}{\eta} \qquad (2)$$

$$I = \frac{P}{V_{bus}} \qquad (3)$$

where F and f_x are the propulsion and friction force, respectively; m is the mass of the person and EW; and v and I are the values of the velocity and current of the EW, respectively. These equations correlated the load/power of the EW motor with the electrical current, as the bus voltage of the SA-HESS almost remains constant. So, the proposed algorithm mainly depended upon electrical current values. Figure 5 shows a flowchart of the proposed algorithm. This methodology was implemented in the Arduino MEGA 2560 interface with MATLAB (R2017a, MathWorks, Natick, MA, USA). The microcontroller, which works as a SEMS, decided the operation mode based on the sensing and analyzing HESS and load parameters (currents, SOC, and voltage). When the EW motor started, most of the power/current was supplied by the SC. V_{SC} and I_L were monitored and analyzed. If the SEMS sensed a small current that was less than the first threshold value (I_{TH1} = 2.5 A), it switched on the LIB circuitry to supply the required current to the load. This condition occurred when the EW was traveling along track-AB. When EW reached point B (see Figure 4b), the SEMS controller sensed a higher value of current compared with I_{TH1} and one lower than a second threshold current (I_{TH2} = 3.5 A). Soon after, SEMS would check V_{SC}. If V_{SC} was higher than the threshold (V_{TH} = 10 V) value, the SC supplied power to the load; otherwise, the LIB and SC supplied power together to the EW motor. Similarly, when EW was at point C (i.e., the current required to the load was more than the I_{TH2}), the controller would switch on both the LIB and SC to fulfill the desired power demands of the load.

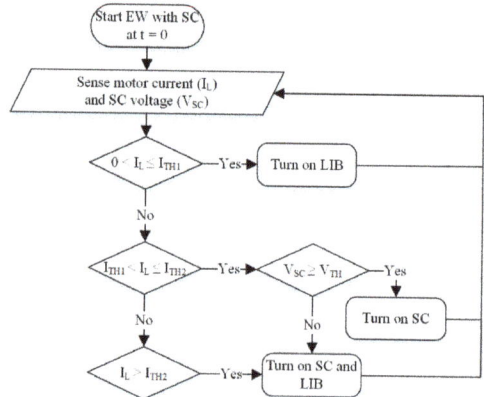

Figure 5. Algorithm for the SEMS.

Figure 6 presents a simplified block of a hardware implementation for the SEMS algorithm. Solid lines represent energy flow lines, and dotted lines show the control signal flow lines. The microcontroller sent a control signal to the switch based on the condition provided by the algorithm and opened a path for the storage system to supply the desired power to the load. Three switches that connected HESS to the EW motor were used: switch SW_1 connected the SC and LIB to the motor, SW_2 was the interface between the SC and load, and SW_3 was between the LIB and load, as shown in Figure 6.

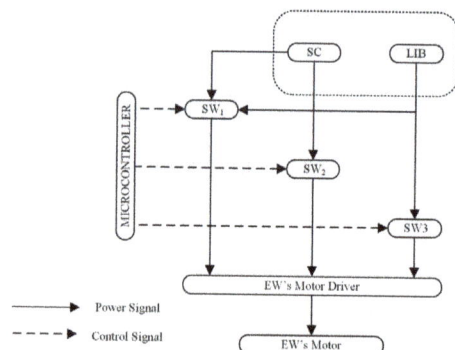

Figure 6. Hardware implementation of the SEMS algorithm.

The voltage requirement and maximum power demand determined the number of series and parallel connected branches of the devices in the storage system. The number of cells in series (N_{bat_s}) and parallel (N_{bat_p}) of the LIB was computed using Equations (4) and (5) [48]:

$$N_{bat\_s} = \frac{V_{bus}}{V_{bat}} \quad (4)$$

$$N_{bat\_p} = \frac{Total\ Ah}{cell\ Ah} \quad (5)$$

Similarly, the number of series and parallel SCs was calculated as [48]

$$N_{SC\_s} = \frac{V_{bus}}{V_{SC}} \qquad (6)$$

$$N_{SC\_p} = \frac{I.N_{SC\_s}}{\Delta V}\left(\frac{\Delta t}{C} + ESR\right) \qquad (7)$$

where N_{SC_s} and N_{SC_p} are the number of supercapacitors connected in series and parallel, respectively, and V_{SC} is the voltage of the supercapacitor. ESR represents the equivalent series resistance, where ΔV and Δt are the voltage drop and discharging time of the supercapacitor depending upon the output current.

3. Experimental Setup

Figure 7 presents the experimental setup of the proposed SA-HESS prototype for an EW. A Samsung 18650 LIB (Samsung, Yongin-si, South Korea) was used in the current experiment. An SC bank with a capacitance of 350 F was used. A liquid crystal display (20 × 4 dimension) was used to indicate the mode of the HESS. The SEMS algorithm implemented in Arduino Mega 2560 was connected to the monitoring system. Different sensors were used to measure and monitor the parameters, such as the HESS temperature, current, and voltage. ACS712-20A hall effect current sensors were used for current sensing. An F30S60S power diode (ON Semiconductor, Phoenix, AZ, USA) was used. A DC motor was used as a load in this prototype.

Figure 7. Experimental setup of the prototype.

4. Results

4.1. Performance Evaluation of Smart Energy Management System

4.1.1. Charging System Using SEMS

Figure 8 shows the voltages and temperature profiles of an LIB pack, where V1, V2, V3, V4, and Vt are the first, second, third, fourth, and overall voltages of the LIB pack, respectively. T1, T2, T3, T4, T5, T6, and T7 denote the value of different temperature sensors. The voltage and temperature were monitored to obtain the optimal value of the charging current for an LIB. At 3498 s, the temperature (T2) increased from the threshold value. To compensate for this effect, the current was reduced, as described in Section 2.2 and shown in the magnified graph.

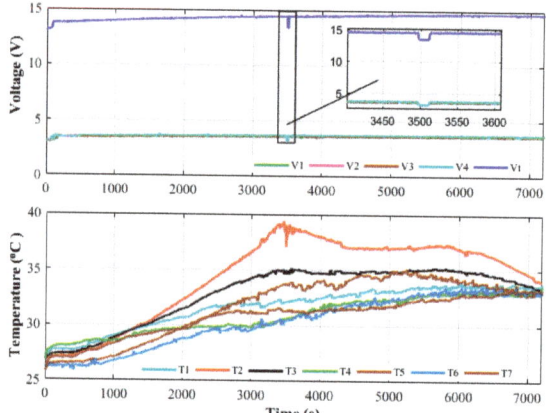

Figure 8. Voltage and temperature profile of the Li-ion battery (LIB).

4.1.2. Discharging System Using SEMS

Two tracks were used to validate the SEMS methodology for SA-HESS, as shown in Figure 9. The EW needed a high current value to start, which was provided by the SC at the time (t) = 9 s. At t = 11 s, the EW started moving on a flat surface. On a normal surface, the EW motor required less current, which was supplied by the LIB at t = 11–50 s. At t = 51 s, the EW started to climb uphill. The EW drew more current, which was provided by the SC. Here, the stress on the LIB was reduced using the SC instead of the LIB. At t = 100 s, the EW reached the middle of the inclined surface and required more power to reach the top surface. In this case, both the LIB and SC supplied power to the load. These timeframes are discussed in detail below in the test cases considering the proposed algorithm, as described in Section 2.3.

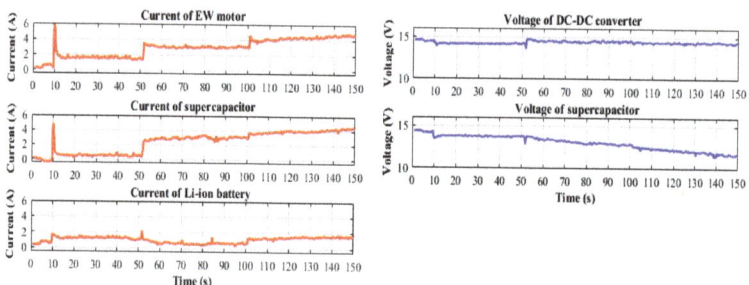

Figure 9. Experimental results of the SEMS on the electric wheelchair (EW) motor.

Case—EW Start (t = 0–11 s)

At the start, the EW motor required a high current for a small amount of time. The current profile in Figure 10 shows a spike at t = 10 s, indicating that the EW motor drew 6 A. The SC supplied this high pulse current. The voltage of the SC showed small variations (0.6 V), but the DC–DC converter maintained a constant voltage for smooth power transfer.

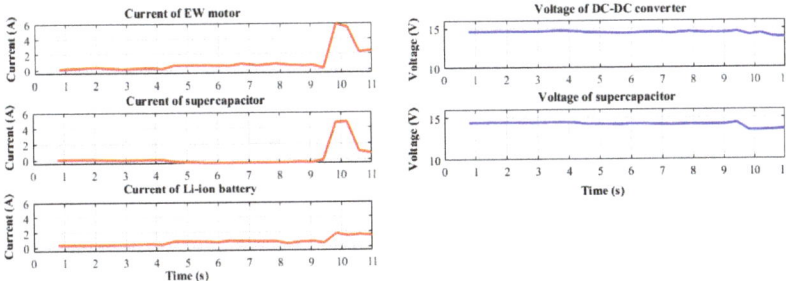

Figure 10. The experimental results at the start of the EW movement.

Case—Plain Track (t = 11–50 s)

From t = 11 to 50 s, the EW traveled on track-AB (from Figure 4a). The EW motor drew a 1.9-A current. SEMS enabled the LIB to supply this small amount of power to the load. The current profile in Figure 11 reveals the SC to have had an almost zero current, while the current of the LIB was 1.3 A. This confirms that only the LIB supplied a small amount of power to the load.

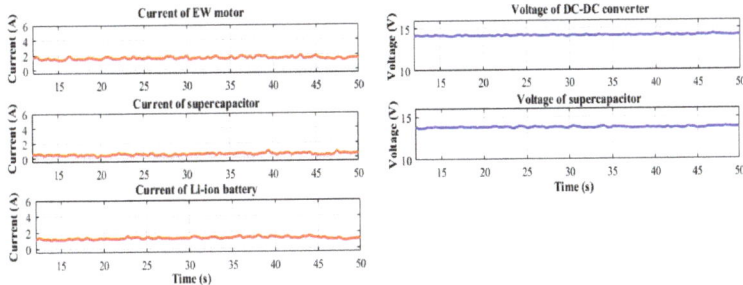

Figure 11. Experimental results for track-AB.

Case—Climbing Uphill (t = 51–150 s)

As shown in Figure 12, from t = 51 to 100 s, the EW traveled from points B to C (as described in Figure 4b). On this track, a 3-A current was drawn by the load. The proposed system enabled SC to supply a high power/current to the load. The SC current increased to 3 A to fulfill the power requirement of the load shown in the current of a supercapacitor in Figure 12. The voltage of the SC started decreasing, but the DC–DC converter maintained a constant voltage to fully utilize the SC. The current profile of the LIB confirmed that the LIB supplied an almost zero current, even at a high load current.

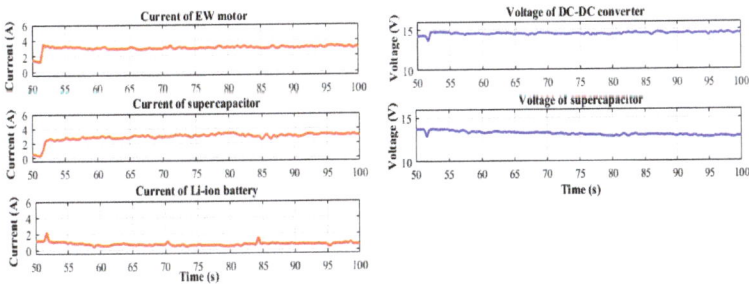

Figure 12. Experimental results for track-ABCD (points B–C).

Figure 13 shows the current and voltage profile of the EW traveling from points C to D from t = 100 to 150 s. At the peak load, the proposed system supplied power from both sources. The experimental results showed that at point C, the LIB supplied approximately 1–1.7 A and the SC supplied 4 A, as shown in the current profile of the LIB and SC in Figure 13. The terminal voltage of the SC decreased from 12.9 to 11.9 V, but the DC–DC converter provided a constant voltage to the load. Regardless of acceleration, the proposed system supplied a constant LIB voltage to the load and less current, which improved the LIB lifecycle. This should also increase the traveling range of the EW.

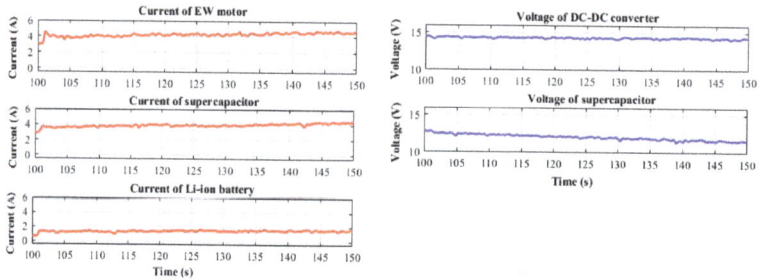

Figure 13. Experimental results for track-ABCD (points C–D).

4.2. Performance Evaluation of P-HESS

The same tests were performed on a P-HESS for track-ABCD. Figure 14 shows the experimental results of the P-HESS. At t = 11 s, the EW motor started, and an approximately 3.5-A current was supplied by the LIB and 3.1 A was supplied by the SC, as shown in the current profiles of the LIB and SC, respectively, in Figure 14. From t = 12 to 50 s, the passive system traveled on a flat surface. The EW motor drew an average of 1.9 A, which was supplied by the LIB, whereas the SC current was almost zero. From t = 51 to 101 s, when it started traveling on an inclined surface, it drew 4.5 A. The LIB and SC supplied 3- and 1.5-A currents, respectively. Similarly, at t = 101–150 s, the passive system traveling on an inclined surface required more current, which was again supplied by the LIB. The current profile of the EW motor increased from 5 to 6 A. The LIB current also increased from 4 to 5 A. The SC supplied 1 A, as shown in the graph of the current of a supercapacitor (Figure 14).

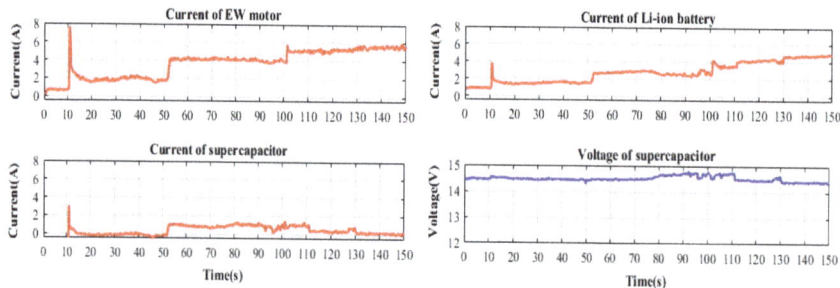

Figure 14. Experimental results of the passive hybrid energy storage system (P-HESS).

5. Discussion

To validate the proposed technique, several experiments were performed to compare the result of SA-HESS and P-HESS under different conditions. The average current values of different track experiments are presented in Figure 15. It can be noted that by adopting the proposed technique, the stress of the LIB was reduced. In high-power-demanding conditions, the proposed technique effectively used the SC as compared with P-HESS to enhance the lifecycle of the LIB. Some of the results of real-time testing under different tracks are shown in Figure A1 of Appendix A.

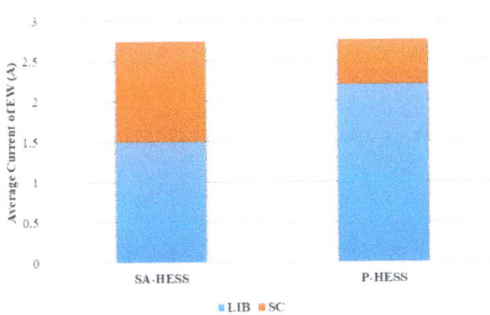

Figure 15. Comparison of the proposed semiactive (SA)-HESS and P-HESS.

The validation of the proposed system was also done under different load conditions. Figure 16 shows some of the experimental results of the proposed system and the P-HESS under different loading conditions. The superiority of the proposed algorithm under different loading conditions can be seen in Figure 16. The efficiency of the SA-HESS was 97.6% in the EW application. The proposed methodology does not have any significant effect on the cost of the system.

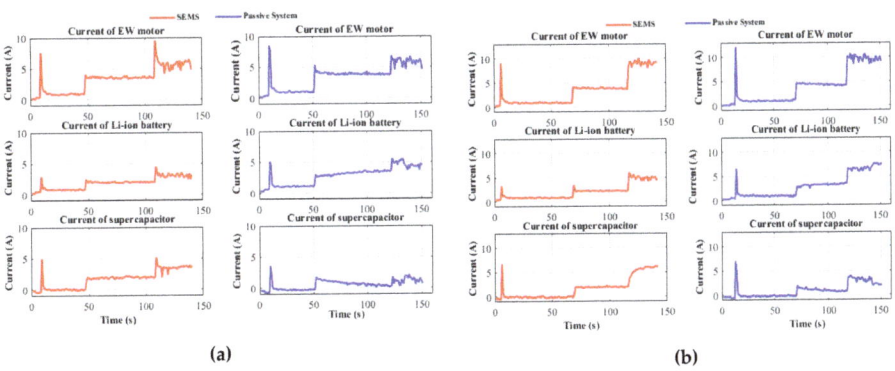

Figure 16. Comparison of the proposed SA-HESS and P-HESS during different loads: (**a**) 35-W load; (**b**) 25-W load.

Renewable energy resources can be used to charge the HESS. The installation of a regenerative braking system in EWs is a good option to charge the SC during the downslope. The discharging efficiency of the HESS can be enhanced by using a smart controller such as an FLC.

6. Conclusions

This paper introduced a new hybrid energy management system for EWs. The SA-HESS was implemented using a smart energy management system algorithm. The proposed system ensured effective use of the SC and decreased the stress of the LIB to extend the battery life under demanding conditions. Five different tracks and two different loads were used to ensure the practicability of the proposed hybrid energy management system and to compare it with the passive system. The experimental results confirmed that the proposed system provided a more effective charging and discharging management system at high power demand compared with the conventional P-HESS.

Author Contributions: Conceptualization, S.H., M.U.A., and H.-J.K.; Formal analysis, S.H., S.H.N., and M.I.; Funding acquisition, H.-J.K.; Investigation, S.H.; Methodology, S.H. and M.U.A.; Software, M.U.A. and I.K.; Supervision, H.-J.K.; Validation, M.U.A., S.H.N., and M.I.; Writing—original draft, S.H., S.H.N., and I.K.; Writing—review & editing, H.-J.K.

Funding: This research was supported by Brain Korea 21 Center for Creative Human Resource Development Program for IT Convergence of Pusan National University.

Acknowledgments: This research was supported by Brain Korea 21 Center for Creative Human Resource Development Program for IT Convergence of Pusan National University.

Conflicts of Interest: The authors declare no conflict of interests.

Appendix A

Figure A1 shows the current profiles of the proposed system and P-HESS while traveling in different surface conditions.

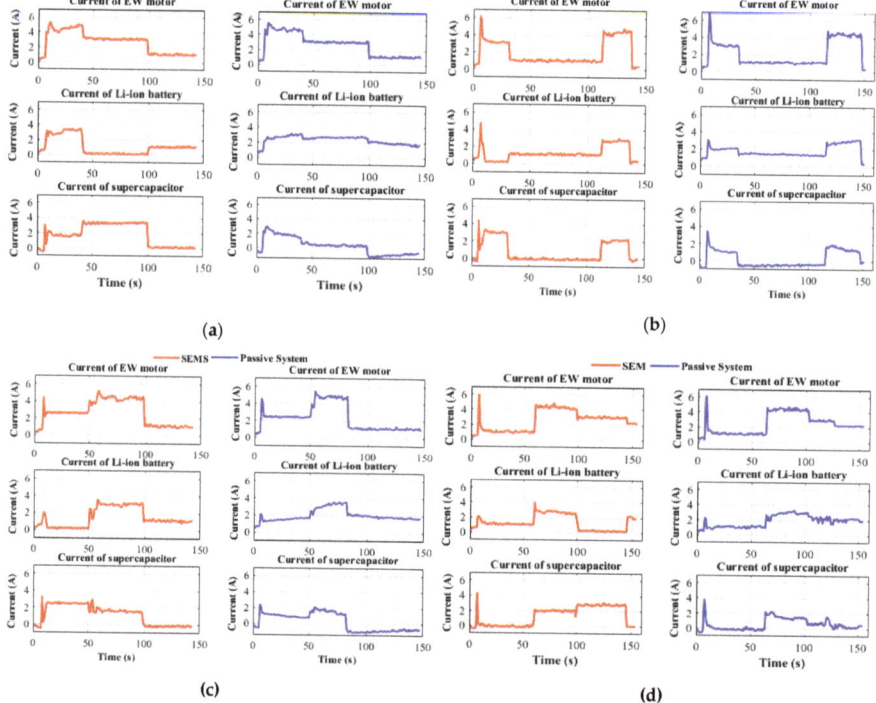

Figure A1. Comparison of the proposed SEMS and P-HESS during different modes: (**a**) start–high climbing–low climbing–plain; (**b**) start–low climbing–plain–high climbing; (**c**) start–low climbing–high climbing–plain; (**d**) start–plain–high climbing–low climbing.

References

1. Tian, Z.; Xu, W. Electric wheelchair controller based on parameter self-adjusting fuzzy PID. In Proceedings of the International Conference on Computational Intelligence and Natural Computing, CINC'09, Wuhan, China, 6–7 June 2009; pp. 358–361.
2. Yang, Y.-P.; Guan, R.-M.; Huang, Y.-M. Hybrid fuel cell powertrain for a powered wheelchair driven by rim motors. *J. Power Sources* **2012**, *212*, 192–204. [CrossRef]

3. Khan, M.A.; Zeb, K.; Sathishkumar, P.; Ali, M.U.; Uddin, W.; Hussain, S.; Ishfaq, M.; Khan, I.; Cho, H.-G.; Kim, H.-J. A Novel Supercapacitor/Lithium Ion Hybrid Energy System with a Fuzzy Logic-Controlled Fast Charging and Intelligent Energy Management System. *Electronics* **2018**, *7*, 63. [CrossRef]
4. Cooper, R.A.; Cooper, R.; Boninger, M.L. Trends and issues in wheelchair technologies. *Assist. Technol.* **2008**, *20*, 61–72. [CrossRef] [PubMed]
5. Cortés, U.; Annicchiarico, R.; Vázquez-Salceda, J.; Urdiales, C.; Cañamero, L.; López, M.; Sànchez-Marrè, M.; Caltagirone, C. Assistive technologies for the disabled and for the new generation of senior citizens: The e-Tools architecture. *AI Commun.* **2003**, *16*, 193–207.
6. Cano, Z.P.; Banham, D.; Ye, S.; Hintennach, A.; Lu, J.; Fowler, M.; Chen, Z. Batteries and fuel cells for emerging electric vehicle markets. *Nat. Energy* **2018**, *3*, 279. [CrossRef]
7. Ding, D.; Cooper, R.A. Electric powered wheelchairs. *IEEE Control Syst. Mag.* **2005**, *25*, 22–34.
8. Ding, Y.; Cano, Z.P.; Yu, A.; Lu, J.; Chen, Z. Automotive Li-Ion Batteries: Current Status and Future Perspectives. *Electrochem. Energy Rev.* **2019**, *2*, 1–28. [CrossRef]
9. Kim, T.; Song, W.; Son, D.-Y.; Ono, L.K.; Qi, Y. Lithium-ion batteries: Outlook on present, future, and hybridized technologies. *J. Mater. Chem. A* **2019**, *7*, 2942–2964. [CrossRef]
10. Hussain Nengroo, S.; Ali, M.U.; Zafar, A.; Hussain, S.; Murtaza, T.; Junaid Alvi, M.; Raghavendra, K.V.G.; Jee Kim, H. An Optimized Methodology for a Hybrid Photo-Voltaic and Energy Storage System Connected to a Low-Voltage Grid. *Electronics* **2019**, *8*, 176. [CrossRef]
11. Ren, G.; Ma, G.; Cong, N. Review of electrical energy storage system for vehicular applications. *Renew. Sustain. Energy Rev.* **2015**, *41*, 225–236. [CrossRef]
12. Hussain Nengroo, S.; Kamran, M.; Ali, M.U.; Kim, D.-H.; Kim, M.-S.; Hussain, A.; Kim, H. Dual battery storage system: An optimized strategy for the utilization of renewable photovoltaic energy in the United Kingdom. *Electronics* **2018**, *7*, 177. [CrossRef]
13. Masih-Tehrani, M.; Ha'iri-Yazdi, M.-R.; Esfahanian, V.; Safaei, A. Optimum sizing and optimum energy management of a hybrid energy storage system for lithium battery life improvement. *J. Power Sources* **2013**, *244*, 2–10. [CrossRef]
14. Ralon, P.; Taylor, M.; Ilas, A.; Diaz-Bone, H.; Kairies, K. *Electricity Storage and Renewables: Costs and Markets to 2030*; International Renewable Energy Agency: Abu Dhabi, UAE, 2017.
15. Ali, M.U.; Zafar, A.; Hussain Nengroo, S.; Hussain, S.; Alvi, M.J.; Kim, H.-J. Towards a Smarter Battery Management System for Electric Vehicle Applications: A Critical Review of Lithium-Ion Battery State of Charge Estimation. *Energies* **2019**, *12*, 446. [CrossRef]
16. Gao, L.; Dougal, R.A.; Liu, S. Power enhancement of an actively controlled battery/ultracapacitor hybrid. *Ieee Trans. Power Electron.* **2005**, *20*, 236–243. [CrossRef]
17. Di Napoli, A.; Ndokaj, A. Auxiliary power buffer based on ultracapacitors. In Proceedings of the 2012 International Symposium on Power Electronics, Electrical Drives, Automation and Motion (SPEEDAM), Sorrento, Italy, 20–22 June 2012; pp. 759–763.
18. Yin, H.; Zhao, C.; Li, M.; Ma, C. Utility function-based real-time control of a battery ultracapacitor hybrid energy system. *IEEE Trans. Ind. Inform.* **2015**, *11*, 220–231. [CrossRef]
19. Hochgraf, C.G.; Basco, J.K.; Bohn, T.P.; Bloom, I. Effect of ultracapacitor-modified PHEV protocol on performance degradation in lithium-ion cells. *J. Power Sources* **2014**, *246*, 965–969. [CrossRef]
20. Zimmermann, T.; Keil, P.; Hofmann, M.; Horsche, M.F.; Pichlmaier, S.; Jossen, A. Review of system topologies for hybrid electrical energy storage systems. *J. Energy Storage* **2016**, *8*, 78–90. [CrossRef]
21. Miller, J.M. Energy storage technology markets and application's: Ultracapacitors in combination with lithium-ion. In Proceedings of the 7th Internatonal Conference on Power Electronics, ICPE'07, Daegu, Korea, 22–26 October 2007; pp. 16–22.
22. Erdinc, O.; Vural, B.; Uzunoglu, M. A wavelet-fuzzy logic based energy management strategy for a fuel cell/battery/ultra-capacitor hybrid vehicular power system. *J. Power Sources* **2009**, *194*, 369–380. [CrossRef]
23. Zhang, Q.; Deng, W. An Adaptive Energy Management System for Electric Vehicles Based on Driving Cycle Identification and Wavelet Transform. *Energies* **2016**, *9*, 341. [CrossRef]
24. Xiong, R.; Chen, H.; Wang, C.; Sun, F. Towards a smarter hybrid energy storage system based on battery and ultracapacitor-A critical review on topology and energy management. *J. Clean. Prod.* **2018**, *202*, 1228–1240. [CrossRef]

25. Kuperman, A.; Aharon, I. Battery–ultracapacitor hybrids for pulsed current loads: A review. *Renew. Sustain. Energy Rev.* **2011**, *15*, 981–992. [CrossRef]
26. Dougal, R.A.; Liu, S.; White, R.E. Power and life extension of battery-ultracapacitor hybrids. *IEEE Trans. Compon. Packag. Technol.* **2002**, *25*, 120–131. [CrossRef]
27. Chen, Z. High pulse power system through engineering battery-capacitor combination. In Proceedings of the (IECEC) 35th Intersociety Energy Conversion Engineering Conference and Exhibit, Las Vegas, NV, USA, 24–28 July 2000; pp. 752–755.
28. Miller, J. Battery-capacitor power source for digital communication applications: Simulations using advanced electrochemical capacitors. *Electrochem. Soc. Proc.* **1995**, *29–95*, 246–254.
29. Tie, S.F.; Tan, C.W. A review of energy sources and energy management system in electric vehicles. *Renew. Sustain. Energy Rev.* **2013**, *20*, 82–102. [CrossRef]
30. Ju, F.; Zhang, Q.; Deng, W.; Li, J. Review of structures and control of battery-supercapacitor hybrid energy storage system for electric vehicles. In Proceedings of the 2014 IEEE International Conference on Automation Science and Engineering (CASE), Taipei, Taiwan, 18–22 August 2014; pp. 143–148.
31. Song, Z.; Hofmann, H.; Li, J.; Han, X.; Zhang, X.; Ouyang, M. A comparison study of different semi-active hybrid energy storage system topologies for electric vehicles. *J. Power Sources* **2015**, *274*, 400–411. [CrossRef]
32. Cope, R.C.; Podrazhansky, Y. The art of battery charging. In Proceedings of the Fourteenth Annual Battery Conference on Applications and Advances, Long Beach, CA, USA, 12–15 January 1999; pp. 233–235.
33. Ali, M.U.; Kamran, M.; Kumar, P.; Hussain Nengroo, S.; Khan, M.; Hussain, A.; Kim, H.-J. An Online Data-Driven Model Identification and Adaptive State of Charge Estimation Approach for Lithium-ion-Batteries Using the Lagrange Multiplier Method. *Energies* **2018**, *11*, 2940. [CrossRef]
34. Hussein, A.A.-H.; Batarseh, I. A review of charging algorithms for nickel and lithium battery chargers. *IEEE Trans. Veh. Technol.* **2011**, *60*, 830–838. [CrossRef]
35. Tsang, K.; Chan, W. Current sensorless quick charger for lithium-ion batteries. *Energy Convers. Manag.* **2011**, *52*, 1593–1595. [CrossRef]
36. Wei, Z.; Meng, S.; Xiong, B.; Ji, D.; Tseng, K.J. Enhanced online model identification and state of charge estimation for lithium-ion battery with a FBCRLS based observer. *Appl. Energy* **2016**, *181*, 332–341. [CrossRef]
37. Wei, Z.; Zou, C.; Leng, F.; Soong, B.H.; Tseng, K.-J. Online model identification and state-of-charge estimate for lithium-ion battery with a recursive total least squares-based observer. *IEEE Trans. Ind. Electron.* **2018**, *65*, 1336–1346. [CrossRef]
38. Wei, Z.; Zhao, J.; Ji, D.; Tseng, K.J. A multi-timescale estimator for battery state of charge and capacity dual estimation based on an online identified model. *Appl. Energy* **2017**, *204*, 1264–1274. [CrossRef]
39. Pay, S.; Baghzouz, Y. Effectiveness of battery-supercapacitor combination in electric vehicles. In Proceedings of the 2003 IEEE Bologna Power Tech Conference Proceedings, Bologna, Italy, 23–26 June 2003; Volume 3, p. 6.
40. Rashid, M. *Power Electronics Circuits, Devices, and Application*; Ptrentice-Hall International Inc.: London, UK, 1993.
41. Huang, J.-W.; Liu, Y.-H.; Wang, S.-C.; Yang, Z.-Z. Fuzzy-control-based five-step Li-ion battery charger. In Proceedings of the International Conference on Power Electronics and Drive Systems (PEDS 2009), Taipei, Taiwan, 3–6 November 2009; pp. 1547–1551.
42. Wangsupphaphol, A.; Idris, N.; Jusoh, A.; Muhamad, N.; Yao, L.W. The energy management control strategy for electric vehicle applications. In Proceedings of the 2014 International Conference and Utility Exhibition on Green Energy for Sustainable Development (ICUE), Pattaya City, Thailand, 19–21 March 2014; pp. 1–5.
43. Ortúzar, M.; Moreno, J.; Dixon, J. Ultracapacitor-based auxiliary energy system for an electric vehicle: Implementation and evaluation. *IEEE Trans. Ind. Electron.* **2007**, *54*, 2147–2156. [CrossRef]
44. Cui, X.; Shen, W.; Zhang, Y.; Hu, C. A Fast Multi-Switched Inductor Balancing System Based on a Fuzzy Logic Controller for Lithium-Ion Battery Packs in Electric Vehicles. *Energies* **2017**, *10*, 1034. [CrossRef]
45. Ali, M.U.; Hussain Nengroo, S.; Adil Khan, M.; Zeb, K.; Ahmad Kamran, M.; Kim, H.-J. A real-time simulink interfaced fast-charging methodology of lithium-ion batteries under temperature feedback with fuzzy logic control. *Energies* **2018**, *11*, 1122.
46. Azizi, I.; Radjeai, H. A new strategy for battery and supercapacitor energy management for an urban electric vehicle. *Electr. Eng.* **2018**, *100*, 667–676. [CrossRef]

47. Hybrid Wheelchair. Available online: https://mafiadoc.com/hybrid-wheelchair_59c3e8f11723dd285cbc616e.html (accessed on 9 January 2019).
48. Tammineedi, C. Modeling Battery-Ultracapacitor Hybrid Systems for Solar and Wind Applications. Master's Dessertation, The Pennsylvania State University Graduate School, State College, PA, USA, 2011.

© 2019 by the authors. Licensee MDPI, Basel, Switzerland. This article is an open access article distributed under the terms and conditions of the Creative Commons Attribution (CC BY) license (http://creativecommons.org/licenses/by/4.0/).

MDPI
St. Alban-Anlage 66
4052 Basel
Switzerland
Tel. +41 61 683 77 34
Fax +41 61 302 89 18
www.mdpi.com

Electronics Editorial Office
E-mail: electronics@mdpi.com
www.mdpi.com/journal/electronics

www.ingramcontent.com/pod-product-compliance
Lightning Source LLC
LaVergne TN
LVHW070706100526
838202LV00013B/1037